Textbook of Agronomy

Textbook of Agronomy

Cassius Foster

Callisto Reference

www.callistoreference.com

Callisto Reference,
118-35 Queens Blvd., Suite 400,
Forest Hills, NY 11375, USA

Visit us on the World Wide Web at:
www.callistoreference.com

ISBN: 978-1-64116-519-8 (Hardback)

Cataloging-in-Publication Data

Textbook of agronomy / Cassius Foster.
p. cm.
Includes bibliographical references and index.
ISBN 978-1-64116-519-8
1. Agronomy. 2. Crops. 3. Soil management. 4. Agriculture. I. Foster, Cassius.
SB91 .T49 2022
630--dc23

Table of Contents

Preface

Agronomy is a domain of science and technology that is concerned with the production and usage of plants for food, fiber, fuel and land reclamation. It encompasses studies from various areas including plant genetics, plant physiology, meteorology and soil science. It uses a combination of sciences like biology, economics, ecology, chemistry and genetics. It primarily deals with issues such as food production, management of the environmental impact of agriculture and extraction of energy from plants. Some focus areas related to agronomy are crop rotation, soil classification, irrigation and drainage, plant physiology, and pest control. It also makes use of biotechnology for achieving the development of required characteristics in plants. This textbook provides comprehensive insights into the field of agronomy. Some of the diverse topics covered herein address the varied branches that fall under this category. This book will provide comprehensive knowledge to the readers.

To facilitate a deeper understanding of the contents of this book a short introduction of every chapter is written below:

Chapter 1- The science and technology of producing and using plants for food, fiber, land reclamation and fuel is known as agronomy. It is the branch of agriculture that is concerned with the principles and practices of soil management and crop production. This is an introductory chapter which will introduce briefly all the significant aspects of agronomy and agronomic crops.

Chapter 2- Tillage is the agricultural preparation of soil with the help of various ways of mechanical agitation such as stirring, overturning and digging. Primary tillage and secondary tillage are its two main types. The topics elaborated in this chapter will help in gaining a better perspective about the practices of tillage.

Chapter 3- An undesirable plant in a human controlled setting is known as a weed. Its management involves the process of controlling weeds. Some of the different forms of weeds are green foxtail, dodder weed and rag weed. This chapter has been carefully written to provide an easy understanding of the different types of weeds and their management.

Chapter 4- Irrigation is a method that involves the application of controlled amount of water to plants at regular intervals. The main methods of irrigation include drip irrigation, surface irrigation, sprinkler irrigation and subsurface drip irrigation. This chapter discusses in detail the methods and types of irrigation as well as their management.

Chapter 5- The processes which are associated with the production of plant-based crops are known as cultivation practices. There are many types of field crops which are cultivated such as cereals, pulses and fibre crops. The chapter closely examines the common cultivation practices related to these crops to provide an extensive understanding of the subject.

Chapter 6- A community of separate farm systems which have broadly similar enterprise patterns, household livelihoods and resource bases is known as a farming system. There are various areas of study under farming systems such as plantation agriculture, crop rotation, dry farming and ley farming. The following chapter elucidates these varied processes and mechanisms associated with this area of study.

Finally, I would like to thank the entire team involved in the inception of this book for their valuable time and contribution. This book would not have been possible without their efforts. I would also like to thank my friends and family for their constant support.

Cassius Foster

Chapter 1

Introduction to Agronomy

The science and technology of producing and using plants for food, fiber, land reclamation and fuel is known as agronomy. It is the branch of agriculture that is concerned with the principles and practices of soil management and crop production. This is an introductory chapter which will introduce briefly all the significant aspects of agronomy.

Agronomy is the branch of agriculture that deals with field crop production and soil management. Agronomists generally work with crops that are grown on a large scale (e.g., small grains) and that require relatively little management. Agronomic experiments focus on a variety of factors relating to crop plants, including yield, diseases, cultivation, and sensitivity to factors such as climate and soil.

Crop management, and its scientific study agronomy, are part of a system that comprises the physical elements of the climate, soil and land, the biological constituents of the vegetation and soil, the economic opportunities and constraints of markets, sales and profit, and the social circumstances and preferences of those who work the land.

Agronomy

Scope of Agronomy

Agronomy is a dynamic discipline with the advancement of knowledge and better understanding of planet, environment and agriculture. Agronomy science becomes imperative in Agriculture in the following areas:

- Identification of proper season for cultivation of wide range of crops is needed which could be made possible only by Agronomy science.

- Proper methods of cultivation are needed to reduce the cost of cultivation and maximize the yield and economic returns.
- Availability and application of chemical fertilizers has necessitated the generation of knowledge to reduce the ill-effects due to excess application and yield losses due to the unscientific manner of application.
- Availability of herbicides for control of weeds has led to development for a vast knowledge about selectivity, time & method of its application.
- Water management practices play grater role in present day crisis of water demand and Agronomy science answer to the questions 'how much to apply?' and 'when to apply'.
- Intensive cropping is the need of the day and proper time and space intensification not only increase the production but also reduces the environmental hazards.
- New technology to overcome the effect of moisture stress under dry land condition is explored by Agronomy and future agriculture is depends on dry land agriculture.
- Packages of practices to explore full potential of new varieties of crops are the most important aspects in crop production which could be made possible only by Agronomy science.
- Keeping farm implements in good shape and utilizing efficient manner to nullify the present day labour crisis is further broadening the scope of agronomy.
- Maintaining the ecological balance through efficient management of crops, livestock and their feedings in a rational manner is possible only by knowing agronomic principles.
- Care and disposal of farm and animal products like milk and eggs and proper maintenance of accounts of all transactions concerning farm business is governing principles of agronomy. Relation of agronomy.

Relation of Agronomy to other Sciences

Agronomy is a main branch of Agriculture. It is synthesis of several disciplines like soil science, Agricultural chemistry, crop physiology, plant ecology, biochemistry and economics:

- The Soil Science helps the agronomist to thoroughly understand the soil physical, chemical and biological properties to effect modification of the soil environment.

- The Agricultural Chemistry help the agronomist to understand the chemical composition and changes involved in the production, protection, and use of crops and livestock.
- The crop physiology helps to understand the basic life process of crops to understand functioning of each parts of plant to determine their input requirement like nutrients etc.
- The plant ecology helps us to understand the associated environment in which the crops grown like the influence of weather (Temperature, Rainfall etc).
- The biochemistry shows the way in which biochemical process takes place in crops which helps to understand critical requirements to favourably activate this process.
- The economics paves the way for profit and loss analysis in farming.

Agronomic Crops

Food Crops

The term "food crops" refers to the world's major food supply derived from plants; a crop assumes human intervention through agriculture. In the main, food crops consist of grains, legumes (including dried beans), seeds and nuts, vegetables, fruit, herbs and spices, beverage plants such as tea and coffee, and so forth.

Some foods, like sea vegetables, are food crops in terms of gathering rather than deliberate agriculture, although undoubtedly the best sources of such foods are well cared for resources in terms of laws to prevent over-harvesting, polluting, and licensing access.

Grains

Grains are a principal source of human food, with wheat, rice, barley, oats, millet, rye and corn (maize) being the most widely consumed sources of grains around the world. Grains are the seeds of grasses. Buckwheat is mistaken for a grain, but is the seed of a herbaceous plant, not a grass. Wheat, Millet and Rice are the Richest Source of Proteins.

Grains have served as a staple food for much of the world's population for thousands of years. In particular, the staples of rye, wheat, barley and oats have been temperate climate favourites, whilst rice, corn and millet have been favored in the tropics and sub-tropics. In modern times, affluent countries are able to enjoy a very wide range of grains that once used to be very localised. However, affluent countries have also taken to using grain to feed livestock to produce meat and less grains are eaten at the dinner table as staple produce as a result.

Dried Beans and Legumes

These are another staple in many diets around the world. Generally cheap to buy, dried beans and legumes tend to also have a long shelf life, making them good for seasons when food is more scarce. They are a high quality source of protein, although on their own they tend to be less tasty and need to be combined with other foods to make them palatable. Besides protein, they contain fibre, carbohydrate, minerals and vitamins needed by the human body. They don't have vitamin C unless sprouted and eaten as sprouts.

One well known food made from soybeans is tofu or bean curd, which is then used in thousands of different ways in varied dishes. Although tasteless, it takes on other flavours readily, and its texture makes it ideal to add in place of meat in many dishes.

Seeds and Nuts

The variety of seeds and nuts available as food crops is enormous. Both seeds and nuts are botanically the same because they consist of a kernel that contains a future plant in embryonic form. As such, they are a powerhouse of energy and a concentrated form of food.

Vegetables

Vegetables are an excellent source of vitamins, minerals and roughage in the human diet. Leaves are particularly important in terms of nutrients, while root vegetables contain the starch and natural sugars that supply humans with energy.

Vegetables are grown both commercially and as a typical backyard or sustainable lifestyle food crop. Organic vegetables tend to cost more and tend to have higher nutritional content. Some say they also taste better than conventionally grown vegetable crops. not all naturally grown foods are vegetables.

Fruit

Fruit tends to be grown in orchards, berry patches, on vines and even in bogs (for cranberries). Most fruit has a specific season and many fruits spoil quickly after picking, so fast action to get them to market or into refrigeration is essential. Wild sources of fruit are not usually considered to be a food crop, although established seasons of collecting berries from areas known for bountiful crops can become a traditional source of berries in a localised area.

Most fruit contains vitamin C, making it an important source of vitamin C in humans. Dried fruit is another common source of fruit in the diet and parts of fruit crops are often set aside for this purpose. Fruit tends to be grown commercially and in home gardens, although fruit trees can take many years to be sufficiently mature to produce adequate fruit for a household.

Herbs and Spices

Herbs are mostly annuals but there are biannuals and perennials too. Herbs consist mainly of the best leaves, stems, flowers, seeds and roots. Some herbs are very strong in flavour. It is commonplace to grow fresh herbs for sale in supermarkets in affluent countries, as an option for a kitchen garden and live, fresh herbs to cook with.

Spices consist of the dried roots, bark, pods, berries or seeds of aromatic plants. Examples include nutmeg, pepper, cinnamon and saffron. Much of the history of trade is bound up in the transfer of spices from East to West, as well as the discovery of vanilla, chillies and allspice in the New World.

Feed

Feed crop or animal feed is the food grown or developed for livestock and poultry. Modern feeds are produced by carefully selecting and blending ingredients to provide highly nutritional diets that both maintain the health of the animals and increase the quality of such end products as meat, milk, or eggs. Ongoing improvements in animal diets have resulted from research, experimentation, and chemical analysis by agricultural scientists.

Animals in general require the same nutrients as humans. Some feeds, such as pasture grasses, hay and silage crops, and certain cereal grains, are grown specifically for animals. Other feeds, such as sugar beet pulp, brewers' grains, and pineapple bran, are by-products that remain after a food crop has been processed for human use. Surplus food crops, such as wheat, other cereals, fruits, vegetables, and roots, may also be fed to animals.

Basic Types of Feeds

Animal feeds are classified as follows:

1. Concentrates, high in energy value, including fat, cereal grains and their by-products (barley, corn, oats, rye, wheat), high-protein oil meals or cakes (soybean, canola, cottonseed, peanut [groundnut]), and by-products from processing of sugar beets, sugarcane, animals, and fish.

2. Roughages, including pasture grasses, hays, silage, root crops, straw, and stover (cornstalks).

Concentrate Foods

Cereal Grains and their by-products

In the agricultural practices of North America and northern Europe, barley, corn, oats, rye, and sorghums are grown almost entirely as animal feed, although small quantities are processed for human consumption as well. These grains are fed whole or ground, either singly or mixed with high-protein oil meals or other by-products, minerals, and vitamins to form a complete feed for pigs and poultry or an adequate dietary supplement for ruminants and horses.

The production of grains is seasonal because of temperature or moisture conditions or a combination of both. It is necessary to produce a full year's supply during the limited growing season. The grain is dried to 14 percent or less moisture to prevent sprouting or molding; the grain is then stored in containers or buildings where insects and rodents cannot destroy it. It is generally desirable to store more than a year's supply of the grains to be used as feed, because crop failures sometimes occur.

High-protein Meals

Vegetable seeds produced primarily as a source of oil for human food and industrial uses include soybeans, peanuts (groundnuts), flaxseed (linseed), canola, cottonseed, coconuts, oil palm, and sunflower seeds. After these seeds are processed to remove the oil, the residues, which may contain from 5 percent to less than 1 percent of fat and 20 to 50 percent of protein, are marketed as animal feeds. Cottonseed and peanuts have woody hulls or shells, which are generally removed before processing—if the hulls or shells are left intact, the resulting by-product is higher in fibre and appreciably lower in protein and energy value. The protein quality of these meals for monogastrics varies greatly depending on the levels and availability of the amino acids present. Ruminants in general require only protein or nitrogen sources for the rumen microbes to synthesize amino acids.

These high-protein feeds supplement inexpensive roughages, cereal grains, and other low-protein feeds in order to furnish the protein and amino acids needed for efficient growth or production. The supplement chosen for a particular diet depends largely on the cost and availability of supply.

By-products of Sugar Beets and Sugarcane

From the sugar beet industry come beet tops, which are used on the farm either fresh or ensiled, and dried beet pulp and beet molasses, which are produced in sugar factories. Cane molasses is a residue from cane sugar manufacture. These are all palatable, high-quality sources of carbohydrates. Sugarcane bagasse (stalk residue) is fibrous, hard to digest, and of very low feed value. In Europe, beets and some other roots are grown as animal feed. Citrus molasses and dried citrus pulp, which are generally available at low cost as by-products of the citrus juice industry, are often used as high-quality feeds for cattle and sheep.

Other by-Product Feeds

Large quantities of animal feed are by-products or residues from commercial processing of cereal grains for human consumption. The largest group of these by-product feeds comes from the milling of wheat, including wheat bran, wheat middlings, wheat germ meal, and wheat mill feed. In some areas, bakery wastes, such as stale and leftover bread, rolls, and various pastry products, are ground and used as filler or feed for pets and farm animals. Rice bran and rice hulls are obtained in similar fashion from the mills that polish rice for human food. Corn gluten feed, corn gluten meal, and hominy feed are produced as by-products from the manufacture of starch for industrial and food uses.

Brewers' grains, corn distillers' grains and solubles, and brewer's yeast are useful animal feeds and are collected from the dried residues of the fermentation industries that produce beer and distilled spirits. Waste products from pineapple-canning plants include pineapple bran or pulp and the ensiled leaves from the plant. By-products from the abattoirs and meatpacking plants that process animals into meat include such feeds as meat and bonemeal, tankage (animal residue left after rendering fat in a slaughterhouse), meat scraps, blood meal, poultry waste, and feather meal. Various types and qualities of fish meals are produced by fish-processing plants. These animal by-products typically contain 50 percent or more high-quality protein and the mineral elements calcium and phosphorus. Steamed bonemeal is particularly high in these important minerals. Dried skim milk, dried whey, and dried buttermilk are feed by-products from the dairy industry.

Roughages

Pasture

Pasture grasses and legumes, both native and cultivated, are the most important single source of feed for ruminants such as cattle, horses, sheep, and goats. During the growing season they furnish most of the feed for these animals at a cost lower than for feeds that need to be harvested, processed, and transported. Hundreds of different grasses, legumes, bushes, and trees are acceptable as feeds for grazing animals. The nutritive value of the cultivated varieties has been studied, but information is incomplete for many of those that occur naturally.

Hay

Hay is produced by drying grasses or legumes when they approach the stage of maximum plant growth and before the seed develops. This stage has been shown to give maximum yields of digestible protein and carbohydrates per unit of land area. The moisture content is typically reduced below 18 percent in order to prevent molding, heating, and spoilage during storage. Legume hays, such as alfalfa and clovers, are high in protein, while the grasses (such as timothy and Sudan grass) are lower in protein and vary considerably depending on their stage of maturity and the amount of nitrogen fertilization applied to them. Stored hay is fed to animals when sufficient fresh pasture grass is not available.

Hay

Silage

Silage is made by packing immature plants in an airtight storage container and allowing fermentation to develop acetic and lactic acids, which preserve the moist feed. Storage may be in upright tower silos or in trenches in the ground. The initial moisture concentration of the forage should be between 50 and 70 percent, depending on the type of silage. Lower moisture levels can cause difficulty in obtaining sufficient packing to exclude air and may result in molding or other spoilage. Too high a moisture content causes nutrient losses by seepage and results in the production of excessively acidic, unpalatable silage. Ensiled forage can be stored for a longer period of time with lower loss of nutrients than dry hay. The nutritive value of silage depends on the type of forage ensiled and how successfully it has been cured. Corn, sorghums, grasses, and sometimes leguminous forages are used in making silage.

Making silage for cattle on a farm

Root Crops

Root crops are used less extensively as animal feed than was true in the past, for economic reasons. Beets (mangels), rutabagas, cassava, turnips, and sometimes surplus potatoes are used as feed. Compared with other feeds, root crops are low in dry-matter content and protein; they mostly provide energy.

Straw and Hulls

Quantities of straw remaining after the harvesting of wheat, oats, barley, and rice crops are used as feed for cattle and other ruminants. The straws are low in protein and very high in fibre; digestibility is low. Straw is useful in maintaining mature animals when other feeds are in short supply, but it is too low in nutrition to be a satisfactory feed for extended periods unless supplemented with other feeds that supply the protein, digestible energy, and minerals needed for growth and production. Treatment of straw with alkali markedly increases the digestibility of the cellulose, augmenting its value as a source of energy for animals.

Corncobs, cornstalks, cottonseed hulls, and rice hulls can also be used as sources of fibre in ruminant diets. Rice hulls are lower in value, while the others are similar to straw.

Fiber Crops

Fiber crops are field crops grown for their fibres, which are traditionally used to make paper, cloth, or rope. The fibers may be chemically modified, like in viscose (used to make rayon and cellophane). In recent years materials scientists have begun exploring further use of these fibers in composite materials.

Fiber crops are generally harvestable after a single growing season, as distinct from trees, which are typically grown for many years before being harvested for such materials as wood pulp fiber or lacebark. In specific circumstances, fiber crops can be superior to wood pulp fiber in terms of technical performance, environmental impact or cost.

Process of Fibre Crops

The process involves first getting the fiber from the plant. The type of fiber will determine the process to be used in the extraction process. Microbes are used in the retting process to get bast fibers. The bacteria help to get rid of the soft tissues from the fiber plant. Decortication is used to get the hard fibers as the machines will remove the product from the plant. Ginning process is also used to get soft fibers since machines involved will remove it from the plant.

Sources of Fiber

Different plants are rich sources of fiber. Such plants produce bast fibers, leaf fibers, and seed fibers. Bast fibers are also known as stem skin fibers and are found in plants such as hemp, papyrus, hoop vine, Kenaf, Nettles, among others. Fibers are also found on leaves in plants like Abaca, Sisal, Yucca, Phormium, and Bowstring hemp. Seeds and fruits can be sources of fiber; they are coir (from coconut), cotton, milkweed, and luffa. Bamboo is another rich source of fiber.Such fibers have different dimensions regarding their diameters and length, which is measured in millimeters. For example, the bamboo has a diameter of 14 millimeters and a length of 2.7 millimeters, which is different from that of cotton, which has a diameter of 25.0 millimeters, and a length of 20 millimeters.

Issues in Fiber Crop Production

While innovation has helped ease the difficulties of fiber production, problems are encountered during storage of fiber crops. It is a challenge since rotting of fiber crops destroys them hence the need to prevent it. The crops are usually stored for a long period; sometimes months, hence it is essential to protect against the elements that might cause rot like water. Due to the presence of different types of fiber crops, storage will depend on the type. For example, the storage of seed fibers will be different from those leaf fibers. The season available will also determine the storage, for example, field crops (like sisal) usually are harvested once a year, unlike the tree crops which can be harvested annually. Supplying the industries with this important product means that storage must be done in the right way.

Types of Fibre Crops

Cotton

- Cotton is also known as White gold in America.
- Botanical name: Gossipium spp.
- Family: Malvaceae.
- Origin: Mexico.

Species of Cotton			
Species of Cotton	Plant Height (CM)	Fiber Length (CM)	Area Covered (per cent) of the Total Acreage of Cotton in india
Indian/Old world cotton:			
Gossipium arborium	1.5-2.0	1.25-2.0	29 per cent
Gossipium herbacium	1.0-1.5	1.5-2.50	21 per cent
American/New world cotton:			
Gossipium hirsutum	1.5	1.8-3.0	50 per cent
Gossipium barbadence	1.5	3.6-5.0	Few thousand ha.

- G. barbadence is also known as Sea Island cotton.
- Black cotton soil is best for cotton cultivation.
- Choulfi method is used for square planting of cotton in Maharashtra and in Malva area of Madhya Pradesh.

Comparative Study of Des and American Cotton		
Particulars	Desu Cotton	American Cotton
Seed rate(kg/ha)	10-12	18-20
Fiber colour	Dul and light brown	White creamy
Ginning per cent	36-42	28-32
Oil content(per cent)	26.0	18.0

Varieties of Cotton	
Desi/Old world	G-777
Interspecific	Varalaxmi,DCH-32(hybrid),HB-224,DHB-105
Intraspecific	H-4,6,Savita,Surya(hybrid),JKHY-1
Egyptian cotton	Sujata ,Savin(Non hybrid)
American cotton	MCU-4,5,8,9

- Nutrient management: 100-120:20-30:15-20 (N: P: Zn kg/ha).
- Water management: 80-90 cm water is required full of its life cycle.
- Cotton is irrigated 2-3 times, 1st irrigation – 30-40 days after sowing 2nd irrigation – Bud fo rmation stage 3rd irrigation – Flowering stage.
- Ginning percentages = $\frac{\text{Wt. of lint}}{\text{Wt. of seed cotton}} \times 100$

  For G. hirsutum: 28-32 per cent (American cotton).

  For G. arborium and G. herbacoum: 36-42 per cent.
- Fibre colour of desi cotton is dull or brown colour, while American cotton has creamy white coloured.
- Lint Index = $\frac{\text{Weight of 100 cotton seed}}{100} \times \text{Ginning\%}$
- H_2SO_4 is used for delinting of cotton seed.
- Categorization of staple length of cotton:

  a) < 19.5 mm-short staple.

b) 20-21.5 mm-medium staple.

c) 22-24.5mm-superior medium.

d) 24.5-26.5mm – long staple.

e) > 27 mm – superior.

- Cotton length is measured by Arealo-meter.
- The oil content in desi cotton is 14.6-25.6 per cent, while American cotton has 14-18 per cent.
- 170 kg cotton = 1 bale.
- Bt-cotton is resistant against the pest Helicoverpa (Spotted bollworm), developed by U.S. based seed company 'Monsanto' and registered under the name 'Bollgaurd'.

Jute

- Jute is a parenchymatus fibre.
- Jute crop does not have hybrid seeds.

Jute Species			
Sl.No.	Particular	White Jute	Black Jute/Tosa
1.	Botenical name	Chorchorus capsularis	Chorchorus olitorius
2.	Planting month	Feb-March	April-May
3.	Grown on	Midland & Lowland	Well drained high land
4.	Tolerate to	Water logging	-
5.	Leaf taste	Bitterness-titapata (due to corchorin)	Testeless-mitapata
6.	Fiber colour	White	Yellow to Jute

- Varieties: JRC-321 (Sonali), JRC-212 (Sabuj sona), JRC 7447 (Shyamli), D154, Hybrid C (Padma), KC1 (joydev) etc.
- For fibre purpose – jute is harvested before flowering.
- Ideal stage for harvesting – Small pod stage/initiation of pod formation (135-140 DAS).

Sun-hemp

- Botanical Name: Crotolaria juncia.
- Best and most widely used green manure crop.

- Seed rate-15-20 kg/ha.
- It is most suitable for alkaline and water logged soil.

References

- Agronomy, science: britannica.com, Retrieved 8 February, 2019
- Food-crops: appropedia.org, Retrieved 17 April, 2019
- Roughages, feed-agriculture: britannica.com, Retrieved 12 January, 2019
- Keyword-fiber: eeb.uconn.edu, Retrieved 18 June, 2019
- What-is-a-fiber-crop: worldatlas.com, Retrieved 29 March, 2019
- Major-types-of-fibre-crops, crops: yourarticlelibrary.com, Retrieved 31 May, 2019

Chapter 2

Tillage Practices

Tillage is the agricultural preparation of soil with the help of various ways of mechanical agitation such as stirring, overturning and digging. Primary tillage and secondary tillage are its two main types. The topics elaborated in this chapter will help in gaining a better perspective about the practices of tillage.

Tillage

Tillage is mechanical modification of soil structure. Tillage tools modify soil structure through a wide range of soil:tool interactions (e.g., cutting, milling, crushing, beating, rebound). The outcome of soil:tool interactions varies with respect to both the characteristics of the tillage operation (e.g., action, depth and width of disturbance, timing) and the characteristics of the soil that is being tilled (e.g., texture, structure, moisture, friability, plasticity).

Importance of Tillage

Tillage has been part of most agricultural systems throughout history because tillage can be used to achieve many agronomic objectives:

- Soil Conditioning (modification of soil structure to favor agronomic processes such as soil seed contact, root proliferation, water infiltration, soil warming, etc.).
- Weed/pest suppression (direct termination or disruption of weed/pest life cycles).
- Residue management (movement, orientation or sizing of residues to minimize negative effects of crop/cover crop residues and promote beneficial effects).
- Incorporation/mixing (placement or redistribution of substances such as fertilizers, manures, seeds, residues, sometimes from a less favorable location to a more favorable spatial distribution).
- Segregation (consolidation of rocks, root crops, soil crumb sizes).
- Land Forming (changing the shape of the soil surface – simplest variant is probably leveling; ridging, roughening and furrowing are also examples).

More specific tillage objectives include seed bed formation, stale seed bed formation, compaction alleviation, fracturing of soil crusts, severing/dessication of weeds, maceration of biofumigant cover crops, stimulation of soil biology, and harvesting of root crops.

Objectives of Tillage

- To prepare a desirable soil structure for a deep seed bed or a root bed suitable for different types of crops. A granular structure is desirable to allow rapid infiltration and good retention of rainfall, to provide adequate air capacity and exchange with in the soil and to minimize resistance to root penetration. A good seed bed generally considered to imply finer particles and greater firmness in the vicinity of the seed.
- To control weeds or to remove unwanted crop plants (thinning).
- To manage plant residues: Mixing of trash is desirable from the tilth and decomposition stand points and retention of trash in the top layers reduce erosion.
- To minimize soil erosion: By following counter tillage, listing and proper placement of trash.
- To establish specific surface configurations for planting, irrigating, drainage, harvesting operations etc.
- To incorporate and mix fertilizers, pesticides, soil amendments etc. in to the soil.

Suitable Method for Gently Sloping Fields

Tillage operations for seed bed preparations are classified as: i) Primary tillage ii) Secondary tillage.

Primary Tillage

The initial major soil working operation designed to plough the soil deeply to reduce soil strength, cover plant materials and rearrange aggregates is called primary tillage. The objectives of primary tillage are:

- To reduce soil strength.
- To rearrange aggregates.
- To cover plant materials and burry weeds.
- To kill insects and pests.

The implements used for primary tillage are called as primary tillage implements They include many animal drawn and tractor drawn implements. Animal drawn implements mostly include indigenous ploughs and mould-board ploughs. Tractor drawn implements include mould-board ploughs, disc ploughs, heavy duty disk harrows, subsoil ploughs, chisel ploughs and other similar implements.

Secondary Tillage

Lighter and finer tillage operations performed in the soil after primary tillage to create proper soil tilth and. surface configuration for seeding and planting are called secondary tillage operations. Secondary tillage operations are generally done on the surface soil. They do not cause much soil inversion and shifting of soil from one place to other. They consume less power per unit area compared to primary tillage operations. The main objectives of secondary tillage are:

- To break the big clods and make the soil surface uniform and leveled as needed for a seed bed.
- To destroy grasses and weeds in the field.
- To cut crop residues and mix them with top soil.

The implements used for secondary tillage operations are called secondary tillage implements. They include different types of harrow, cultivators, sweeps, clod crushers, levellers, bund formers, ridge ploughs etc.

Tillage Systems

Tillage system consists of sequences of operations that manipulate the soil in order to produce a crop. The operations include tilling, planting, fertilization, pesticide application, harvesting, and residue chopping or shredding. The ways in which these operations are implemented affect the physical and chemical properties of the soil, which in turn affect plant growth.

There are two types of tillage systems namely:

- Conservation tillage system including no-till, ridge till and mulch till systems and
- Other than conservation tillage systems namely:
    a. Reduced tillage system.
    b. Conventional tillage system.

Conservation Tillage

Any tillage and planting system that covers 30 percent or more of the soil surface with crop residue, after planting, to reduce soil erosion by water or any system that maintains at least 1,120 kilogram per hectare of flat, small grain residue equivalent on the surface throughout the critical wind erosion period is called conservation tillage system Conversion tillage systems are further classified as a) no-till, b) ridge-till, and c) mulch-till systems.

1. No-till: No-till is defined as a system in which the soil is left undisturbed from harvest to planting except for nutrient injection. Planting or drilling is accomplished in a narrow seedbed or slot created by coulters, row cleaners, disk openers, in-row chisels, or rototillers. Weed control is accomplished primarily with herbicides. Cultivation may be used for emergency weed control.
2. Ridge-till: In ridge-till, the soil is also left undisturbed from harvest to planting except for nutrient injection. Planting is completed in a seedbed prepared on ridges with sweeps, disk openers, coulters, or row cleaners. Residue is left on the surface between ridges. Weed control is accomplished with herbicides and cultivation. Ridges are rebuilt during cultivation.
3. Mulch-till: The soil is disturbed before planting. Tillage tools such as chisels, field cultivators, disks, sweeps or blades are used. Weed control is accomplished with herbicides and/or cultivation. Mulch-till is a category that includes all conservation tillage practices other than no-till and ridge-till. Two tillage practices that fall into this category are zone-till and strip-till. Both of these tillage practices involve tilling a strip into which seed and fertilizer are placed.

Other Tillage Systems

Tillage systems that leave less than 30 percent crop residue after planting are not classified as conservation tillage. However, these systems may meet erosion control goals with or without other supporting conservation practices, such as strip cropping, contouring, terracing, etc.

1. Reduced-till: Reduced-till systems leave 15-30 percent residue cover after planting or 560 to 1,120 kilograms per hectare of small grain residue equivalent throughout the critical wind erosion period.
2. Conventional-till: Conventional-till systems leave less than 15 percent residue cover after planting, or less than 560 kilograms per hectare of small grain residue equivalent throughout the critical wind erosion period. These systems generally involve plowing or some other form of intensive tillage.

Another Classification of Tillage Systems

There are two types of tillage namely: a) Conventional tillage or clean tillage and b) Conservation tillage.

1. Conventional tillage or clean tillage: Ploughing the entire field several times to prepare a seed bed is called conventional tillage.

2. Conservation tillage: Ploughing the field with lesser number of passes over the entire land or ploughing only in the required space of the land and then sowing is called conservation tillage. Different types of conservation tillage are as fallows.

   a. Minimum Tillage: Minimum soil manipulation necessary to meet tillage requirements.

   b. Mulch Tillage: Tillage operations in which nearly 30 % of crop residue or other mulching materials are left on or near the soil surface is called mulch tillage.

Mulch tillage

   c. Rotary: Tillage for crop production is called minimum tillage.

   d. Strip Tillage: In strip tillage system only isolated bands of soil are tilled.- Tillage operations employing rotary action of the tool to cut, break and mix the soil is called rotary tillage.

Strip tillage in wheat

Strip tillage in corn

   e. Combined Tillage: Tillage operations utilizing simultaneously two or more different types of tillage tools or implements to simplify, control or reduce the number of operations over a field is called combined tillage.

Strip till in wheat

Classification of Tillage

Primary Tillage

Primary tillage is the first soil tillage after the last harvest. It is normally conducted when the soil is wet enough to allow plowing and strong enough to give reasonable levels of traction. This can be immediately after the crop harvest or at the beginning of the next wet season. When there is sufficient power available some soil types are ploughed dry.

The objectives of primary tillage are:

- To attain a reasonable depth (10-15 cm) of soft soil with varying clod sizes.
- Kill weeds by burying or cutting and exposing the roots.
- Soil aeration and water accumulation; depending on the soil type and the plough the soil will normally be inverted aerating the deep layers and trapping water during a rainfall event.
- Chop and incorporate crop residues.

Types of Primary Tillage

Depending upon the purpose or necessity, different types of tillage are carried out. They are deep ploughing, subsoiling and year-round tillage.

Deep Tillage

Deep ploughing turns out large sized clods, which are baked by the hot sun when it is done in summer. These clods crumble due to alternate heating and cooling and due to occasional summer showers. This process of gradual disintegration of clods improves soil structure. The rhizomes and tubers of perennial weeds (world's problematic weeds viz., Cynodon dactylon and Cyperus rotundus) die due to exposure to hot sun. Summer deep ploughing kills pests due to exposure of pupae to hot sun.

A deep tillage of 25-30 cm depth is necessary for deep rooted crop like pigeonpea while moderate deep tillage of 15-20 cm is required for maize. Deep tillage also improves soil moisture content. However the advantage of deep tillage in dry farming condition depends on rainfall pattern and crop. It is advisable to go for deep ploughing only for long duration, deep rooted crops. Depth of ploughing should be related to the amount of rainfall that it can wet.

Subsoiling

Hard pans may be present in the soil which restrict root growth of crops. These may be silt pans, iron or aluminium pans, clay pans or -man-made pans. Man-made pans are tillage pans induced by repeated tillage at the same depth. Root growth of crops is confined to top few centimetres of soil where deep penetration of roots is inhibited by hard pans.

For example, cotton roots grow to a depth of 2 m in deep alluvial soil without any pans. When hard pans are present, they grow only up to hard pan, say 15-20 cm. Similarly, vertical root growth of sugarcane is restricted due to hard pans and it is not compensated by horizontal spread. Subsoiling is breaking the hard pan without inversion and with less disturbance of top soil. A narrow cut is made in the top soil while share of the subsoiler shatters hard pans. Chisel ploughs are also used to break hard pans present even at 60-70 cm. The effect of subsoiling does not last long. To avoid closing of subsoil furrow, vertical mulching is adopted.

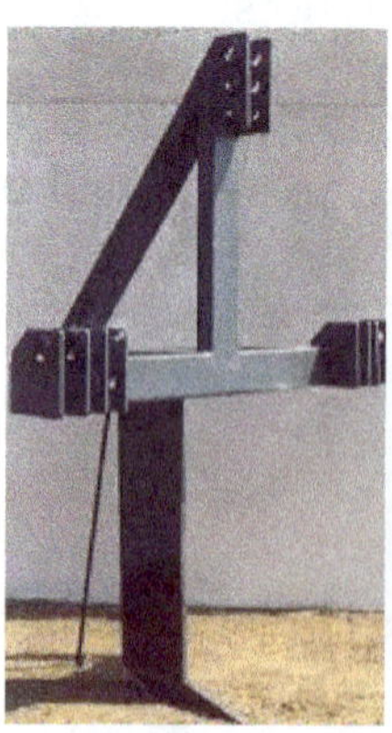

Year-round Tillage

Tillage operations carried out throughout the year are known as year-round tillage. In dry farming regions, field preparation is initiated with the help of summer showers. Repeated tillage operations are carried out until sowing of the crop. Even after harvest of the crop, the field is repeatedly ploughed or harrowed to avoid weed growth in the off season.

Implements for Primary Tillage

Plough - Ploughing is the primary tillage operations, which are performed to cut, break and invert the soil partially or completely. Ploughing essentially means opening the upper crust of the soil, breaking the clods and making the soil suitable for sowing seeds.

Country or Indigenous Plough

It penetrates into the soil and breaks it open. The functional components include share, body, shoe, handle and beam. It can be used for dry land, garden land and wetland ploughing operations:

1. Share - It is the working part of the plough attached to the shoe with which it penetrates into the soil and breaks it open.
2. Shoe - It supports and stabilizes the plough at the required depth.
3. Body - It is main part of the plough to which the shoe, beam and handle are generally attached. In country plough body and shoe are integral part.
4. Beam - It is generally a long wooden piece, which connects the main body of the plough to the yoke.
5. Handle - A wooden piece vertically attached to the body to enable the operator to control the plough.

Operational Adjustments

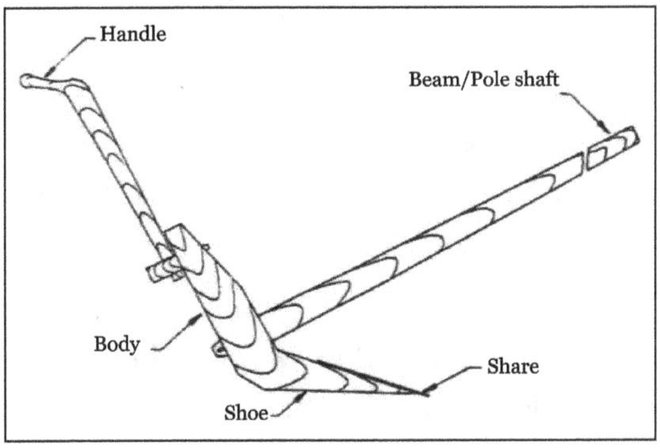

Country Plough

1. Lowering or raising the beam with respect to the plough body, resulting in a change in the angle of the share with the horizontal plane to increase or decrease the depth of operation.

2. Changing the length of the beam (body to yoke on the beam) to increase or decrease the depth of operation.
3. The size of the plough is represented by the width of the body.

Mould Board Plough

Function: 1) cutting the furrow slice, 2) lifting the soil, 3) turning the furrow slice and 4) pulverising the soil.

Components

M.B. plough consists of Share, Mould board, Land side, Frog and Tail piece:

1. Share - It penetrates into the soil and makes a horizontal cut below the soil surface. It is a sharp, well polished and pointed component. The shares are made of chilled cast iron or steel. The steel mainly contains about 0.70 to 0.80% carbon and about 0.50 to 0.80% manganese besides other minor elements.

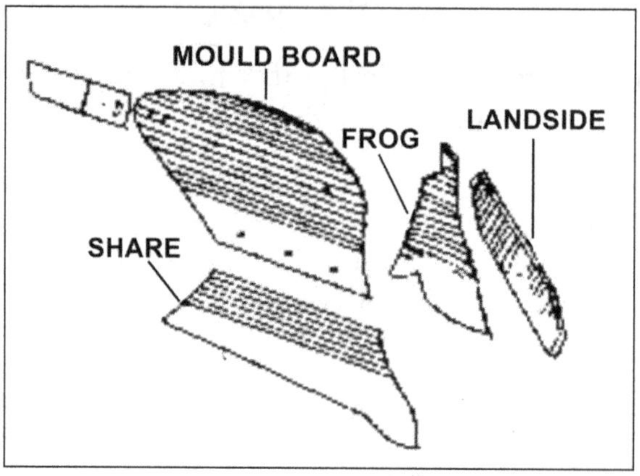

Components of Mould Board plough

2. Mouldboard - The mouldboard is that part of the plough which receives the furrow slice from the share. If lifts, turns and breaks the furrow slice. To suit different soil conditions and crop requirements, mouldboard has been designed in different shapes. The mouldboardis of following types: a) General purpose, b) Stubble, c) Sod or Breaker and d) Slat.

- General purpose - It is a mould board having medium curvature lying between stubble and sod. The sloping of the surface is gradual. It turns the well defined furrow slice and pulverises the soil thoroughly. It has a fairly long mouldboard with a gradual twist, the surface being slightly convex.
- Stubble type -It is short but broader mouldboard with a relatively abrupt curvature which lifts breaks and turns the furrow slice used in stubble soils. Its curvature is not gradual but it is abrupt along the top edge. This causes the furrow

slice to be thrown off quickly, pulverising it much better than other types of mouldboard. This is best suited to work in stubble soil that is under cultivation for years together. Stubble soil is that soil in which stubble of the plants from the previous crop is still left on the land at the time of ploughing.

Types of mould board

- Sod or Breaker type - It is a long mouldboard with gentle curvature which lifts and inverts the unbroken furrow slice. It is used in tough soil of grasses. It turns over thickly covered soil. This is very useful where complete inversion of soil is required by the farmer.
- Slat type - It is a mouldboard whose surface is made of slats placed along the length of the mouldboard, so that there are gaps between the slats. This type of mouldboard is often used, where the soil is sticky, because the solid mouldboard does not scour well in sticky soils:

3. Land side - It is the flat plate which bears against and transmits lateral thrust of the plough bottom to the furrow wall. It helps to resist the side pressure exerted by the furrow slice on the mouldboard. It also helps in stabilizing the plough while it is in operations.

4. Frog - Frog is that part of the plough bottom to which the other components of the plough bottom are attached. It is an irregular piece of metal. It may be made of cast iron for cast iron ploughs or it may be welded steel for steel ploughs.

5. Tail piece - It is an important extension of mouldboard which helps in turning a furrow slice.

Plough accessories There are a few accessories necessary for plough such as:

1. Jointer,
2. Coulter,
3. Gauge wheel,
4. Land wheel and
5. Furrow wheel.

- Jointer - It is a small irregular piece of metal having a shape similar to an ordinary plough bottom. It looks like a miniature plough. Its purpose is to turn over a small ribbon like furrow slice directly in front of the main plough bottom. This small furrow slice is cut from the left and upper side of the main furrow slice and is inverted so that all trashes on the top of the soil are completely turned down and buried under the right hand corner of the furrow.

- Coulter - It is a device used to cut the furrow slice vertically from the land ahead of the plough bottom. It cuts the furrow slice from the land and leaves a clear wall. It also cuts trashes which are covered under the soil by the plough. The coulter may be (a) Rolling type disc coulter or (b) Sliding type knife coulter.
- Gauge wheel - It is an auxiliary wheel of an implement to maintain uniform depth of working. Gauge wheel helps to maintain uniformity in respect of depth of ploughing in different soil conditions. It is usually placed in hanging position.
- Land wheel - It is the wheel of the plough, which runs on the ploughed land.
- Front furrow wheel - It is the front wheel of the plough, which runs in the furrow.
- Rear furrow wheel - It is the rear wheel of the plough, which runs in the furrow.

Adjustment of Mouldboard Plough

Vertical Suction (Vertical Clearance)

It is the maximum clearance under the land side and the horizontal surface when the plough is resting on a horizontal surface in the working position. It is the vertical distance from the ground, measured at the joining point of share and land side. It helps the plough to penetrate into the soil to a proper depth. This clearance varies according to the size of the plough.

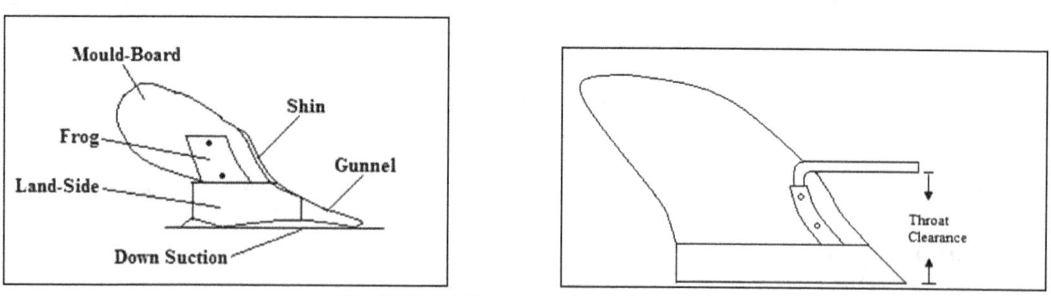

Vertical and horizontal suction

Horizontal Suction (Horizontal Clearance)

It is the maximum clearance between the land side and a horizontal plant touching point of share at its gunnel side and heal of land side. This suction helps the plough to cut the proper width of furrow slice. This clearance varies according to the size of the plough. It is also known as side clearance.

Throat Clearance

It is the perpendicular distance between point of share and lower position of the beam of the plough.

Plough Size

The size of the mouldboard plough is expressed by width of cut of the soil.

Disc Plough: It is a plough, which cuts, turns and in some cases breaks furrow slices by means of separately mounted large steel discs. A disc plough is designed with a view of reduce friction by making a rolling plough bottom instead of sliding plough bottom. A disc plough works well in the conditions where mould board plough does not work satisfactorily.

Advantages of Disc Plough

- A disc plough can be forced to penetrate into the soil which is too hard and dry for working with a mould board plough.
- It works well in sticky soil in which a mould board plough does not scour.
- It is more useful for deep ploughing.
- It can be used safely in stony and stumpy soil without much danger of breakage.
- A disc plough works well even after a considerable part of the disc is worn off in abrasive soil.
- It works in loose soil also (such as peat) without much clogging.

Disadvantages of Disc Plough

- It is not suitable for covering surface trash and weeds as effectively as mouldboard plough does.
- Comparatively, the disc plough leaves the soil in rough and more cloddy condition than that of mouldboard plough.
- Disc plough is much heavier than mouldboard plough for equal capacities because penetration of this plough is affected largely by its weight rather than suction. There is one significant difference between mouldboard plough and disc plough i.e. mouldboard plough is forced into the ground by the suction of the plough, while the disc plough is forced into the ground by its own weight.

Types of Disc Plough

Disc ploughs are of two types (i) Standard disc plough and (ii) Vertical disc plough.

Standard Disc Plough

It consists of steel disc of 60 to 90 cm diameter, set at a certain angle to the direction of travel. Each disc revolves on a stub axle in a thrust bearing, carried at the lower end of a strong stand which is bolted to the plough beam. The angle of the disc to the vertical and to the furrow wall is adjustable. In action, the disc cuts the soil, breaks it and pushes it sideways. There is little inversion of furrow slice as well as little burying of weeds and trashes. The disc plough may be mounted type or trailed type. In mounted disc plough, the side thrust is taken by the wheels of the tractor. Disc is made of heat treated steel of 5 mm to 10 mm thickness. The amount of concavity varies with the diameter of the disc. The approximate values being 8 cm for 60 cm diameter disc and 16 cm for 95 cm diameter. A few important terms connected with disc plough is explained below.

Standard disc plough

1. Disc - It is a circular, concave revolving steel plate used for cutting and inverting the soil.
2. Disc angle - It is the angle at which the plane of the cutting edge of the disc is inclined to the direction of travel. Usually the disc angle of good plough varies between 42° to 45°.
3. Tilt angle - It is the angle at which the plane of the cutting edge of the disc is inclined to a vertical line. The tilt angle varies from 15° to 25° for a good plough.
4. Scraper - It is a device to remove soil that tend to stick to the working surface of a disc.

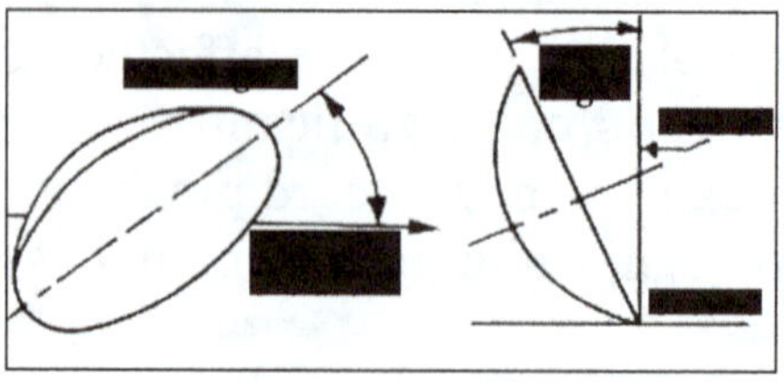

Angles of disc plough

5. Concavity - It is the depth measured at the centre of the disc by placing its concave side on a flat surface.

Vertical Disc Plough

It is the plough which combines the principle of regular disc plough and disc harrow and is used for shallow working in the soil.

Draft of Disc Plough

The disc plough is lighter in draft than the mouldboard plough, turning same volume of soil in similar conditions. In very hard soil, some extra weight is added to the wheel which increases the draft.

Rotary Tiller

The rotary cultivator is widely considered to be the most important tool as it provides fine degree of pulverization enabling the necessary rapid and intimate mixing of soil besides reduction in traction demanded by the tractor driving wheels due to the ability of the soil working blades to provide some forward thrust to the cultivating outfit. Rotary tiller is directly mounted to the tractor with the help of three point linkage. The power is transmitted from the tractor PTO (Power Take Off) shaft to a bevel gear box mounted on the top of the unit, through telescopic shaft and universal joint. From the bevel gear box the drive is further transmitted to a power shaft, chain and sprocket transmission system to the rotor. The tynes are fixed to the rotor and the rotor with tynes revolves in the same direction as the tractor wheels. The number of tynes varies from 28 - 54. A levelling board is attached to the rear side of the unit for levelling the tilled soil. A depth control lever with depth wheel provided on either side of the unit ensures proper depth control. The following types of blades are used with the rotor.

1. 'L' type blade - Works well in trashy conditions, they are more effective in cutting weeds and they do not pulverize the soil much.
2. Twisted blade - Suitable for deep tillage in relatively clean ground, but clogging and wrapping of trashes on the tynes and shafts.
3. Straight blade - Employed on mulchers designed mainly for secondary tillage.

Chisel Plough

Chisel ploughs are used to break through and shatter compacted or otherwise impermeable soil layers. Deep tillage shatters compacted sub soil layers and aids in better infiltration and storage of rainwater in the crop root zone. The improved soil structure also results in better development of root system and the yield of crops and their drought tolerance is also improved. The functional component of the unit include reversible share, tyne (chisel), beam, cross shaft and top link connection.

Sub-soil Plough

The function of the sub-soiler is to penetrate deeper than the conventional cultivation machinery and break up the layers of the soil, which have become compacted due to the movement of heavy machinery or as a result of continuous ploughing at a constant depth. These compacted areas prevent the natural drainage of the soil and also inhibit the passage of air and nutrients through the soil structure. The sub-soiler consists of heavier tyne than the chisel plough to break through impervious layer shattering the sub-soil to a depth of 45 to 75 cm and requires 60 to 100 hp to operate it. The advantages are same as that of chisel plough.

Ploughing System

1. Normal Ploughing: It is the ploughing up to a depth of about 15 cm.
2. Contour Ploughing: It is the method of ploughing in which the soil broken and turned along the contours.

Ploughing of Land

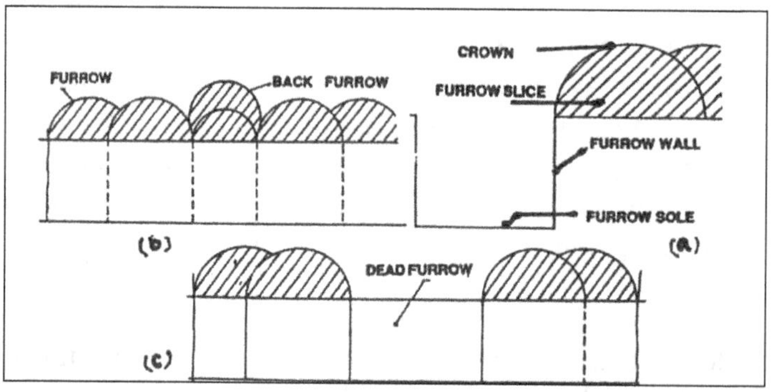

Plough furrow

The ploughing of land separates the top layer of soil into furrow slices. The furrows are turned sideways and inverted to a varying degree, depending upon the type of plough being used. It is a primary tillage operation, which is performed to shatter soil uniformly with partial or complete soil inversion. There are a few important terms frequently used in connection with ploughing of land.

1. Furrow: It is a trench formed by an implement in the soil during the field operation.
2. Furrow slice: The mass of soil cut, lifted and thrown to one side is called furrow slice.

3. Furrow wall: It is an undisturbed soil surface by the side of a furrow.
4. Crown: The top portion of the turned furrow slice is called crown.
5. Back furrow: A raised ridge left at the centre of the strip of land when ploughing is started from centre to side is called back furrow. When the ploughing is started in the middle of a field, furrow is collected across the field and while returning trip another furrow slice is lapped over the first furrow. This is the raised ridge which is named as back furrow.
6. Dead furrow: An open trench left in between two adjacent strips of land after finishing the ploughing is called dead furrow.
7. Head land: While ploughing with a tractor to turn, a strip of un ploughed land is left at each end of the field for the tractor to turn, that is called head land. At the end of each trop, the plough is lifted until the tractor and the plough have turned and are in position to start the return trip. The head land is about 6 metres for two or three bottom tractor plough and one metre more for each additional furrow.

Methods of Ploughing

In order to provide furrows at all times on the right hand side of the plough two method of working are used a) Gathering b) Casting.

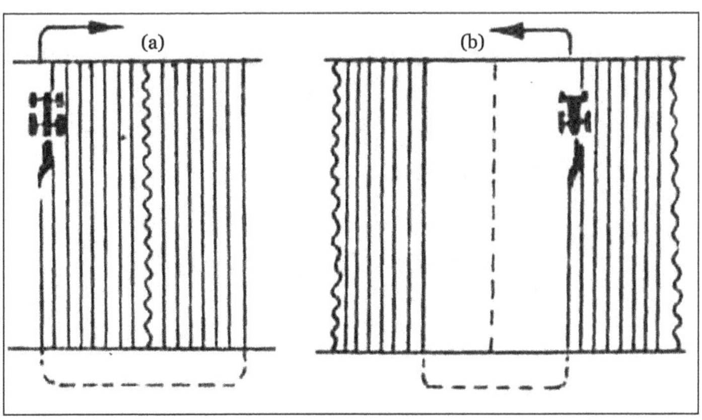

Figure: a. Gathering b. Casting

1. Gathering - Whenever a plough works round a strip of ploughed land, it is said to be gathering.

2. Casting - Whenever a plough works round a strip of unploughed land, it is said to be casting.

Ploughing of a field by casting or gathering alone is normally uneconomical. The following are a few important methods used in tractor ploughing.

1. Continuous ploughing method.
2. Round and round ploughing.

Terminology of Plough

Centre of Power

It is the true point of hitch of a tractor.

Centre of Resistance

It is the point at which the resultant of all the horizontal and vertical forces act. The centre lies at a distance equal to 3/4th size of the plough from the share wing.

Line of Pull

It is an imaginary straight line passing from the centre of resistance through the clevis to the centre of pull (power).

Pull

It is the total force required to pull an implement.

Draft

It is the horizontal component of the pull, parallel to the line of motion.

$D = P \cos \theta$ where D is draft (kgf) and P = pull in (kgf)

θ = angle between line of pull and horizontal.

$$HP = \frac{Draft(kgf) \times Speed(m/s)}{75}$$

Draft depends upon:

1. Sharpness of cutting edge,
2. Working speed,
3. Working width,
4. Working depth,
5. Type of implement,
6. Soil condition and
7. Attachments.

Side Draft

It is the horizontal component of the pull perpendicular to the direction of motion. This is developed if the centre of resistance is not directly behind the centre of pull.

Unit Draft

It is the draft per unit cross sectional area of the furrow.

Theoretical Field Capacity

It is the rate of field coverage of the implement, based on 100 per cent of time at the rated speed and covering 100 per cent of its rated width.

$$TFC, ha/h = \frac{w \times s}{10}$$

Effective Field Capacity

It is the actual area covered by the implement based on its total time consumed and its width.

$$c = \frac{s \times w}{10} \times \frac{E}{100}$$

Where

C = effective field capacity, hectare per h.

S = speed of travel in km per hour.

W = theoretical width of cut of the machine in metre, and

E = field efficiency in percent.

Field Efficiency

It is the ratio of effective field capacity and theoretical field capacity expressed in percent.

$$Field\ Efficiecy = \frac{EFC}{TFC} \times 100$$

Soil Inversion

$$Soil\ inversion = \frac{Number\ of\ weeds\ before\ ploughing - Number\ of\ weeds\ after\ ploughing}{TFC} \times 100$$

Soil Pulverization

It is the quality of work in terms of soil aggregates and clod size. This is measured by cone penetrometer.

Secondary Tillage

Secondary tillage consists of conditioning the soil to meet the different tillage objectives of the farm. These operations consume less power per unit area compared to primary tillage operations. Secondary tillage implements may be tractor drawn or bullock drawn implements.

Harrow

Harrow is a secondary tillage implement that cuts the soil to a shallow depth for smoothening and pulverizing the soil as well as to cut the weeds and to mix the materials with the soil. There are several types of harrow used in India such as disc harrow, spike tooth harrow, spring tooth harrow, acme harrow, patela, triangular harrow, bade harrow, guntaka and reciprocating power harrow.

Disc Harrow

It is a harrow, which performs the harrowing operation by means of a set, or a number of sets of rotating slat discs, each set being mounted on a common shaft. Disc harrow is found very suitable for hard ground, full of stalks and grasses. It cuts the lumps of soil, clods and roots. Disc are mounted on one, two or more axles which may be set at a variable angle to the line of motion. As the harrow is pulled ahead, the discs rotate on the ground. Depending upon the disc arrangements, disc harrows are divided into two classes a) Single action and b) Double action.

Single Action Disc Harrow

It is a harrow with two gangs placed end to end, which throw the soil in opposite directions. The discs are arranged in such a way that right side gang throws the soil towards right, and left side gang throws the soil towards left.

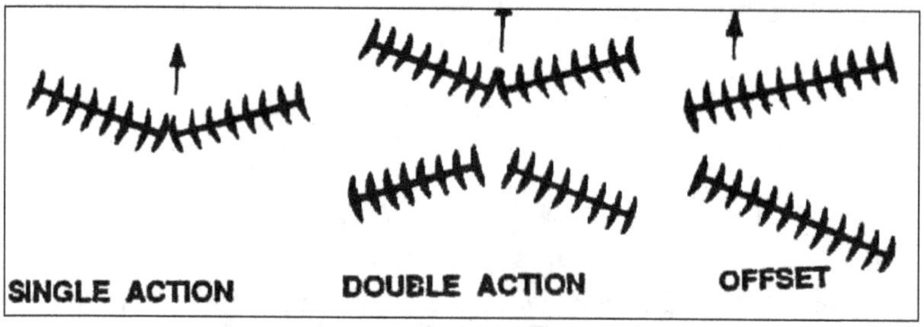

Types of disc harrow

Double Action Disc Harrow

A disc harrow consisting of two or more gangs, in which a set of one or two gangs follow behind the set of the other one or two, arranged in such a way that the front and back gangs throw the soil in opposite directions. Thus the entire field is worked twice in each trip. It may be of two types a) Tandem and b) Off-set.

1. Tandem disc harrow - It is a disc harrow comprising of four gangs in which each gang can be angled in opposite direction.
2. Off-set disc harrow- It is a disc harrow with two gangs in tandem, capable of being offset to either side of the centre line of pull. Two gangs are fitted one behind the other. the soil is thrown in both directions because discs of both gangs face in opposite directions.

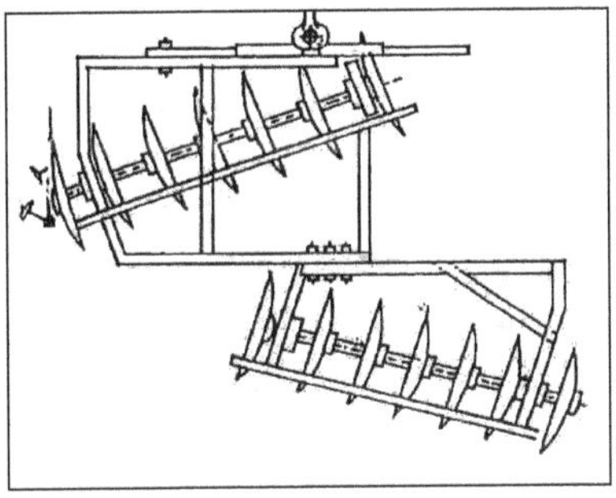

Tandem disc harrow

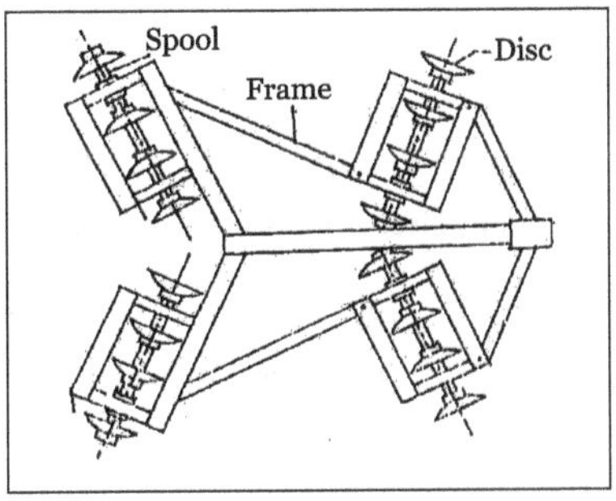

Offset disc harrow

A disc harrow mainly consists of disc, gang, gang bolt, gang central lever, spool or arbor bolt or spacer, bearings, transport wheels, scraper and weight box:

1. Disc: It is a circular concave revolving steel plate used for cutting a inverting the soil. Disc is made of high glass heat-treated hardened steel. Tractor drawn disc harrows have concave discs of size varying from 35-70 cm diameter. Concavity of the disc affects penetration and pulverization of soil. Usually two types of disc are used in disc harrows, plain disc and cut away disc.
2. Gang: Each set of discs that are mounted on a common shaft is called the gang.
3. Gang bolt or arbor bolt: It is a long heavy of square headed bolt from the other end, a set of discs are mounted on the gang bolt. The spacing between the discs on the gang bolt ranges from 15 to 25 cm for light duty and 25 to 30 cm for heavy-duty harrows.
4. Gang control lever: A lever, which operates the gang mechanisms of the disc harrow, is called the gang control lever.
5. Spool or spacer: The flanked tube, mounted on the gang bolt between every

two discs to prevent the lateral movement of the disc on the shaft is called the "spool' or 'spacer'.

6. Bearing: Bearing is essential to counter act the end thrust of the gang due to soil thrust. The harrow bearings are subjected to heavy radial and thrust roods chilled cast iron bearings are used to heavy radial and thrust loads and they are also used due to their durability.

7. Transport wheel: In trailing type disc harrow, the transport wheels are provided for transport work on roads and for preventing the edges of the discs from damage. Mounted type disc harrows do not require wheels for transport work.

8. Scraper: It prevents disc from clogging. It removes the soil that may stick to the concave side of the disc.

9. Weight box: A box like frame is provided on the main frame of the harrow for putting additional weight on the implement. Additional weight helps in increasing the penetration of the disc in the soil.

A sharp edged disc has more effective penetration compared to blunt edged disc. It is observed that penetration is better in low speed than in high speed. The following are a few adjustments for obtaining higher penetration:

1. By increasing the disc angle.
2. By adding additional weight in harrow.
3. By lowering the hitch point.
4. By using the sharp edged discs of small diameter and losses concavity.
5. By regulating the optimum speed.

Spike Tooth Harrow

It is a harrow with peg shaped teeth of diamond cross section to a rectangular frame. It is used to break the clod, stir the soil, uproot the weeds, level the ground, break the soil and cover the seeds. Its principle is to smoothen and level the soil directly after ploughing. Spike tooth harrows may be of rigid type and flexible type. Tractor drawn harrows are usually flexible type. It has got the advantage of being turned up for transporting purpose.This harrow mainly consists of teeth, tooth bar frame, clamps, guard, braces, levers and hooks.

Spring Tooth Harrow

It is a harrow with tough flexible teeth, suitable to work in hard and stony soils. Spring tooth harrow is fitted with springs having loops of elliptical shape. It gives a spring action in working condition. It is used in the soil when obstruction like stone, roots and weeds are hidden below the ground surface. This type pulverizes the soil and helps in killing weeds. The levers are provided for setting the teeth for varying the depth of harrowing. For light harrowing, the adjustment is done in slanting position. Draft hooks on each corner of every section for hitching purpose.

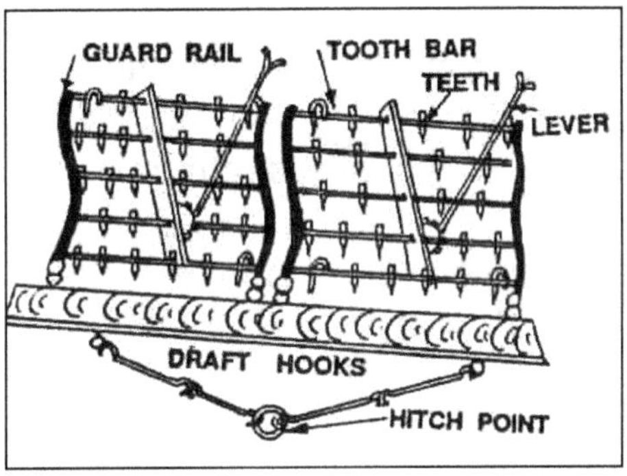

Spike tooth harrow

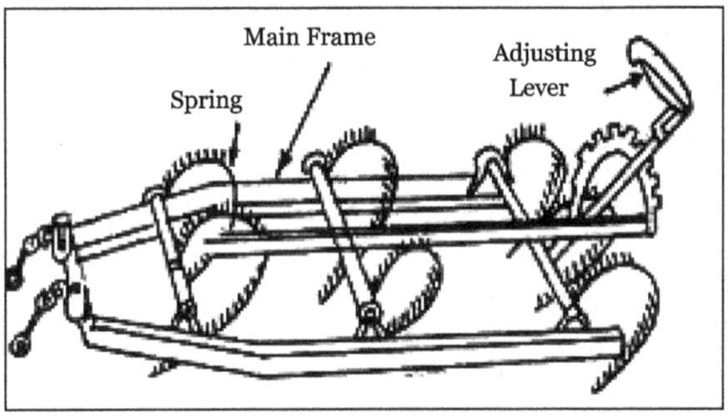

Spring tooth harrow

Acme Harrow

It is a special type of harrow having curved knives. It is also called as knife harrow. The front part of the knife breaks the soil and crushes the clods. This harrow obtains a good pulverization. It is good for mulching also.

Patela

It is a wooden plank used for smoothening the soil and crushing the weeds. It is also used for breaking clods, packing and levelling the ploughed soil and to remove the weeds. It is made of a wooden plank with a number of curved steel hooks bolted to a steel angle section, which are fixed or hinged to the rear side of the plank. The cutting edge levels and packs the soil and the curved hooks uproot and collect the weeds.

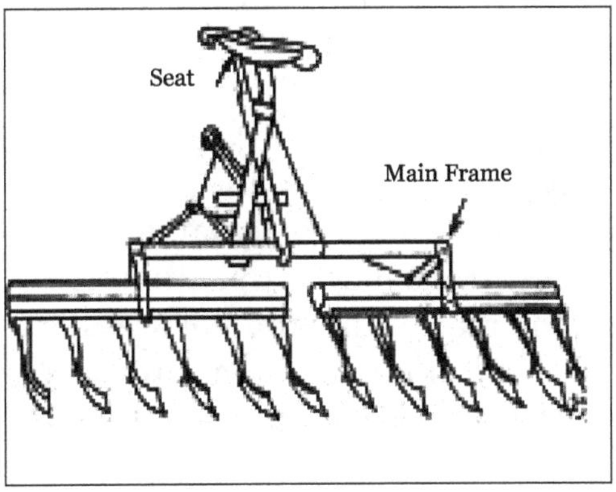

Acme harrow

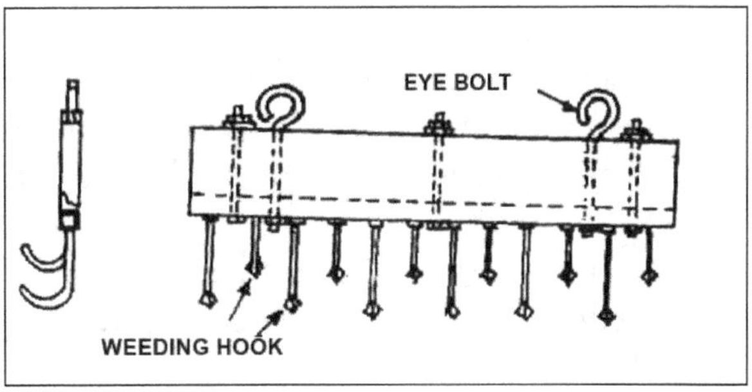

Patela

Triangular Harrow

It is a spike tooth harrow with triangular frame. The frame is made of wood and pointed spikes are fitted in the frame. The teeth of the spikes are fixed and not adjustable.

Blade Harrow (Bakhar)

It is an implement, which consists of one or more blades attached to the beam or frame, used for shallow working of the soil with minimum soil inversion. It is used to prepare seedbeds mostly in clayey soils. It works like a sweep, which moves into top surface of the soil without inverting the soil.

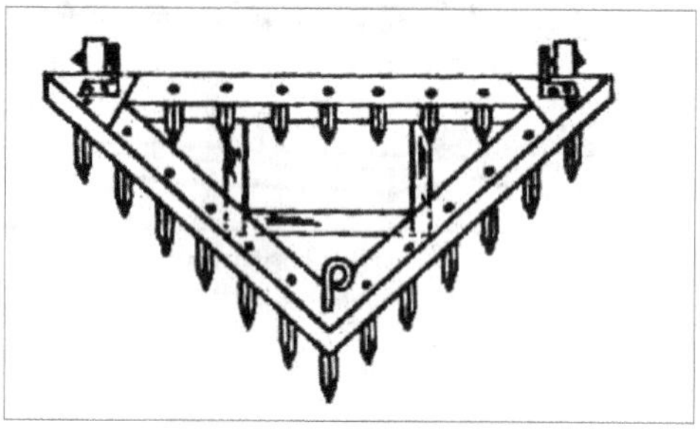

Tri angular harrow

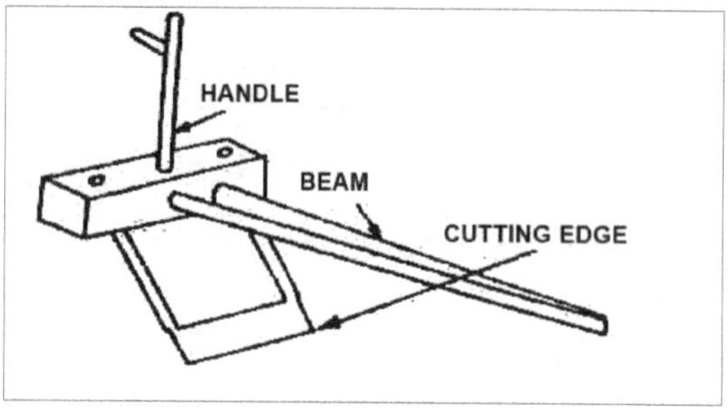

Blade harrow

Guntaka

It is an important type of blade harrow. It is an implement, which consists of one or more blades attached to a frame or beam. It is used for shallow working of the soil with minimum soil inversion. It is mainly used to prepare the seed bed mostly in clay soils. The function of Bakhar the same as that of the guntaka.

Reciprocating Power Harrow

It is a harrow fitted with rigid tynes driven by the power take off in a reciprocating, transverse or rotary motion as the machine moves forward. The power tiller harrow is a rear mounted reciprocatory comb type. It has two horizontal oscillating arms having staggered pegs in two rows at 200 mm spacing. Two sets of slider crank mechanisms provide power from power tiller PTO through universal joint and bevel gear reduction box to the arms. An eccentric embedded flywheel actuates the pitman. The two arms move in opposite directions and hence the implement is dynamically balanced. The amplitude of vibration is 200 mm and the frequency of operation is 400 cycles per minute. A transport wheel provided in the rear of the unit ensures depth control. The unit is attached to the power tiller through the hitch point provided in front portion of the unit. The oscillating pegs breaks the clod, pulverizes the soil to a fine tilth.

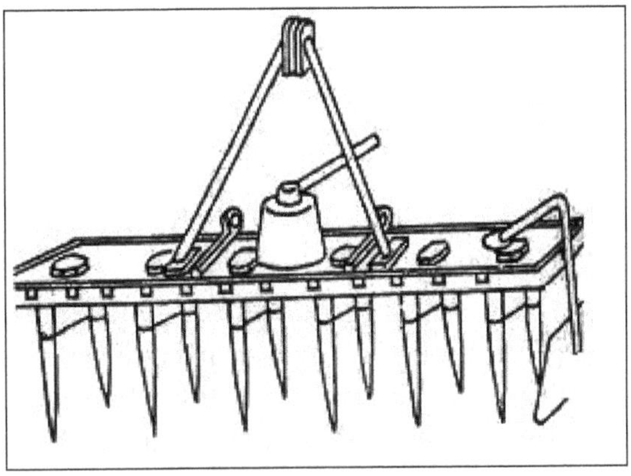

Reciprocating Harrow

Bund Former

It is used for making bunds or ridges by collecting the soil. Bunds are required to hold water in the soil, thereby one can conserve moisture and prevent run-off. The size of the bund former is determined by measuring the maximum horizontal distance between the two rear ends of the farming boards. Bund former consists of forming board, beam and handle.

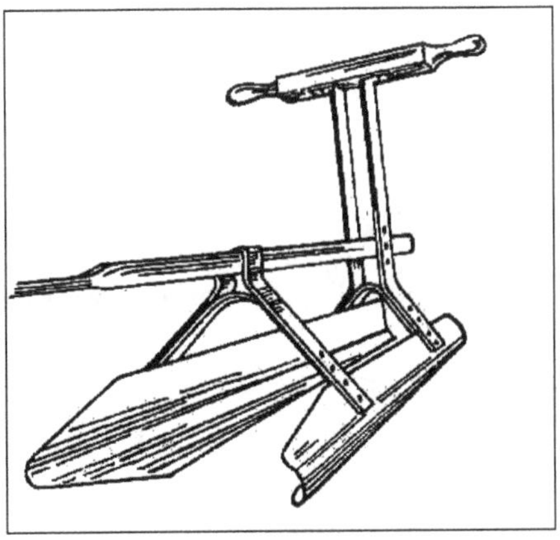

Bund former

Ridger

It is an implement importantly used to form ridges required for sowing row crop seeds and plants in well-tilled soil. The ridger is also used for forming field or channels, earthing up and similar other operations. Ridger is also known as ridging plough and double mould board plough. The ridger generally has 'V' shaped or wedge shaped share fitted to the frog. The nose or tip of share penetrates into the soil and breaks the earth. The mould boards lift, invert and also cast aside the soil, forming deep channels and ridges of the required size. A ridger consists of beam, clevis, frog, handle, mould boards, share and sliding share.

Puddler

Puddler is an implement for churning the soil with water. It is used to prepare paddy fields with standing water after initial ploughing with country plough. It breaks up the clods and churns the soil. The main purpose of puddling is to reduce leaching of water or decrease percolation of water, to kill the weeds by decomposition and to facilitate transplantation of paddy seedlings by making the soil softer. Puddling is done in standing water of 5-10 cm depth. A common puddler has puddling units each having four straight blades or paddles or fan type blades or squirrel type blade mounted on an axle. The axle with the puddling units is fully mounted on two bearings fitted on a frame made of metal or wood. The weight of the puddler is 30-40 kg. A puddler consists of a frame, puddling unit, axle, metal cross beam and handle.

Leveller

Land levelling is expected to bring permanent improvement in the value of land. Levelling work is carried out to modify the existing contours of land so as to achieve certain objectives desired for efficient agricultural production system. These objectives include:

1. Efficient application of irrigation water,
2. Improved surface drainage,
3. Minimum soil erosion,
4. Increased conservation of rain water specially on dry lands, and
5. Provision of an adequate field size and even topography for efficient mechanisation.

Cultivators

It is an implement for inter cultivation with laterally adjustable tines or discs to work between crop rows. The cultivator stirs the soil, and breaks the clods. The tines fitted on the frame of the cultivator comb the soil deeply in the field. A cultivator performs functions intermediate between those of plough and the harrow. Destruction of weeds is the primary function of a cultivator. The following are a few important functions performed by a cultivator.

1. Interculture the fields.
2. Destroy the weeds in the field.
3. Aerate the soil for proper growth of crops.
4. Conserve moisture by preparing mulch on the surface.

5. To sow seeds when it is provided with sowing attachments.
6. To prevent surface evaporation and encourage rapid infiltration of rain water into the soil.

The cultivator can be 1) Disc cultivator, 2) Rotary cultivator, 3) Tine cultivator.

1. Disc cultivator : It is a cultivator fitted with discs.
2. Rotary cultivator : It is a cultivator with tines or blades mounted on a power driven horizontal shaft.
3. Tine cultivator : It is a cultivator fitted with tines having shovels.

Tractor Drawn Cultivator

Trailed Type Cultivator

It consists of a main frame which carries a number of cross members to which tines are fitted. At the forward end of the cultivator, there is a hitch arrangement for hitching purpose. A pair of wheels are provided in the cultivator. The height of the hitch is adjusted so that main frame remains horizontal over a range of depth setting. The tines in each row are spaced widely to allow free passage of the soil and trash around them. The tines in subsequent rows are staggered so that the implement can cover the entire width nicely.

Mounted Cultivator

Tractors fitted with hydraulic lift operate the mounted type cultivators. A rectangular frame of angle iron is mounted on three point hydraulic linkage of the tractor. The cross members carry the tines in two staggered lines. Depending upon the type of soil and crop, shovels are chosen for use on the cultivators. Usually tractor drawn cultivators are of two types, depending upon the flexibility and rigidity of tines (i) Cultivator with spring loaded tines (ii) Cultivator with rigid tynes.

Cultivator with Spring Loaded Tines

A tine hinged to the frame and loaded with a spring so that it swings back when an obstacle is encountered, is called ,spring loaded line. Each of the tine of this type of cultivator is provided with two heavy coil springs, pre-tensioned to ensure minimum movement except when an obstacle is encountered. The springs operate, when the points strike roots or large stones by allowing the tines to ride over the obstruction, thus preventing damage. On passing over the obstruction, the tines are automatically reset and work continues without interruption. The tines are made of high carbon steel and are held in proper alignment on the main frame members. This type of cultivator is particularly recommended for soils which are embedded with stones or stumps.

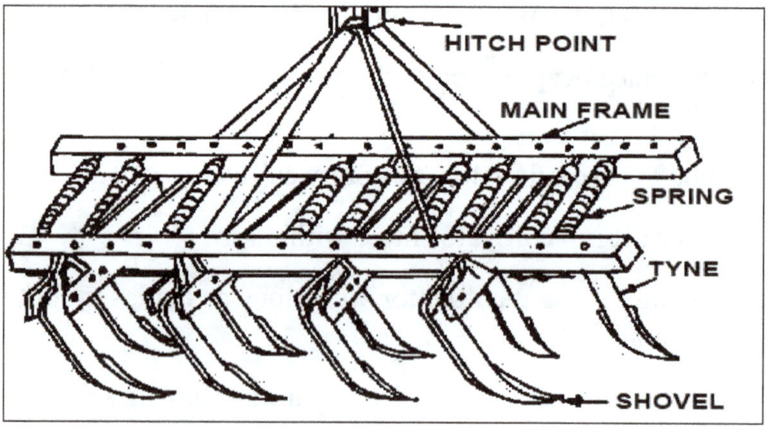

Tractor drawn cultivator

Cultivator with Rigid Tines

Rigid tines of the cultivators are those tines which do not deflect during the work in the field. The tynes are bolted between angle braces, fastened to the main bars by sturdy clamps and bolts . Spacing of the tines is changed simply by slackening the bolts and sliding the braces to the desired position. Since rigid tines are mounted on the front and rear tool bars, the spacing between the tynes can be easily adjusted without getting the tines choked with stubbles of the previous crop or weed growth. A pair of gauge wheel is used for controlling the depth of operation.

Modern Concepts of Tillage

1. The word tillage means to plough and prepare soil for good seed to sow, to cultivate and raise crops.
2. "Tillage is the mechanical manipulation of the soil for obtaining conditions ideal for seed germination, seedling establishment and growth of crops".

3. Jethrotull is the father of tillage.
4. Tilth is the good physical condition of soil after tillage in relation to plant growth.

Types of Tillage

Primary/Properly Tillage

1. The tillage operation which constitutes the initial measure soil working operation. It is normally design to reduce soil strength, cover plant material and rearrange aggregates and mainly for seed bed preparation.
2. The main aim of primary plough is breaking, opening and turning of soil.
3. The primary tillage implements are Desha plough, MB plough. Ridge plough, Disk plough etc.

Secondary Tillage

1. The tillage operation following primary tillage to create a good seedbed for proper seeding/planting.
2. The secondary tillage implements are Cultivator, Harrows, Hoe, Flanker, Roller etc.

Objectives of Tillage

1. Adequate soil aeration for gaseous exchange in the seed and root zone.
2. Adequate seed-soil contact to permit water flows to seed and seedling roots.
3. A non-crust soil contact to permit seedling emergence.
4. A low density soils that permit root elongation and proliferation.
5. An environment that provides adequate light to the seedling (weed free environment).
6. A pest and pathogen free environment.
7. Mixing the applied manures and fertilizers with the soil.
8. Removing the hard pan, if any to increase the soil depth for water absorption.

The aim of the tillage is to produce as good a soil condition or tilth as possible for crop establishment and initial shoot and root development.

Seed-bed

When soil is brought to such a condition ideal for crop growth, it is called seed-bed and is said to be in good tilth. The ideal seed-bed is a soil in which porosity of the mineral solid matter provides an optimum balance between water holding and freely drained. The optimum balance appears to be the one in which capillary pores are in equal proportion.

Conventional Tillage

Conventional tillage involves primary tillage to break, open and turn the soil followed by secondary tillage to obtain seed bed for sowing or planting. With the introduction of herbicides in intensive farming systems, the concept of tillage has undergone radical change:

a. The depth of tillage has been reduced and cultivator replaced the plough as primary tillage implement, particularly on heavy soils.

b. There has been an increasing trend to reduce the number of cultivations in order to avoid undesirable results of over cultivation.

c. Reduction or elimination of tillage has been accompanied by direct drilling of seeds.

d. The greatest success and potential for restricted tillage appear to be on heavy soils, where timeless of cultivation is so critical.

e. The main disadvantages are that direct drilling favors proliferation of perennial weeds and certain pests and diseases. Nutrient leaching may be greater than on ploughed land. Response of direct drilled crops to fertilizers is lesser because of greater weed competition or slower crop establishment. Soil is invariably more compacted and proportion of large pores decreases.

Minimum Tillage

Minimum tillage refers to growing a crop using only that tillage needed for placing the seed at proper depth and then covering them. Thus minimum tillage aimed that reducing tillage to the minimum necessary for ensuring a good seed-bed, rapid germination, satisfactory stands and favorable growing condition.

Tillage can be Reduced in two ways:

a. By omitting operations which do not give much benefits when compared to the cost, and

b. By combining agricultural operations like seeding and tillage.

Advantages of Minimum Tillage

a. Improves soil conditions due to decomposition of plant residues insitu,

b. Improves soil physical condition due to incorporation of organic residues leading to higher water infiltration into the subsoil,

c. Higher infiltration is caused by the vegetation present on the soil and channels formed by the decomposition of dead roots,

d. Less resistant to root growth due to improved soil structure,

e. Lesser surface runoff of rain water,

f. Satisfactory crop stand establishment,

g. Less soil compaction by the reduced movement of heavy tillage vehicles and less erosion compared to conventional tillage.

Disadvantages of Minimum Tillage

a. Seed germination is lower with minimum tillage,

b. In minimum tillage, more nitrogen has to be added as the rate of decomposition of organic matter is slow,

c. Nodulation is affected in some leguminous crop i.e. Field pea, Beans,

d. Sowing operations are difficult with ordinary implements,

e. Continuous use of herbicides causes pollution problem and dominance of perennial problematic weeds.

Methods of Practicing Minimum Tillage

a. Row zone tillage.

b. Plough plant tillage.

c. Wheel track planting.

Row Zone Tillage

Row zone tillage means after primary tillage with MB plough, tillage operations like disking and harrowing are reduced. Tillage is done in row zone only.

Plough Plant Tillage

After the soil is ploughed, a special planter is used and run over the field. The row zones are pulverized and seeds sown.

Wheel Track Planting

Ploughing is done as usual. Tractor is used for sowing and the wheels of tractor pulverize the row zone.

Zero Tillage

Zero tillage referred as no tillage in which the crop is planted in unprepared soil by opening a narrow slot, trench or band only of sufficient width and depth to obtund proper seed coverage. The weeds are controlled only by chemical means:

a. The word 'Zero tillage' was given by Jethrotull.

b. Zero tillage is possible where the soil has: (i) course textured surface, (ii) good internal drainage and (iii) adequate quantity of residue as mulch.

c. Problems of zero tillage are low seed germination, low mineralization and build up of volunteer plants.

d. Commonly used herbicides under zero tillage are parquet and piquet both.

Conservation Tillage

Conservation tillage is a system of tillage in which organic residues are not inverted into the soil such that they remain on the surface as a protective cover against erosion and evaporation losses of soil moisture. If stubbles from the protective cover on surface, it is usually referred as stubble mulch tillage. The residues left on the soil surface interfere with seed-bed preparation and sowing operations.

Effects of Tillage on Plants and Soils

Conservation tillage practices change many soil properties when implemented for a long term. Changes in soil properties change the way in which crops respond to fertilizer management practices. Tillage systems affect soil properties such as temperature, moisture, bulk density, aggregation, organic matter content, and plant properties such as root density.

Effects of Tillage on Soil Temperature Spring Soil

Temperatures are usually lower in conservation tillage systems than in other tillage systems. This results from the insulating effect of residue cover. Residues reflect a portion of the solar energy that would otherwise reach the soil surface. In addition, residues delay soil drying. Higher soil moisture results in cooler spring soil temperatures because wetter soils require more energy to warm than do drier soils.

Moisture Conservation between Tillage Systems

Conservation tillage systems increase the amount of water stored in the soil profile. Surface residues reduce Evaporation and increase water infiltration. The conservation of soil moisture can be very important in regions of reduced rainfall, on soils low in water-holding capacity, and in years with below-average rainfall.

- Soil Bulk Density — Soils in conservation tillage systems generally have higher bulk densities than tilled soils. Bulk density is the weight of soil within a given volume. Soils with higher bulk densities usually have less pore space. This condition can lead to a decreased amount of root growth. In an intensively tilled system, higher bulk densities would result in decreased root growth. However, in conservation tillage systems, higher bulk densities do not necessarily result in reduced root growth. Soils in conservation tillage systems exhibit increased aggregation and higher numbers of root and worm channels than other tillage systems. These channels provide paths for root growth. If the channels are open to the soil surface, water infiltration increases. The differences in bulk density between conservation tillage systems and other tillage systems usually disappears at the end of the growing season, because tilled soils become more dense throughout the season from compaction by rainfall.
- Organic Matter Accumulation — Longterm implementation of conservation tillage practices also increases organic matter levels in the soil. Lower soil temperatures and increased soil moisture contribute to slower rates of organic matter oxidation. Increases in organic matter are normally observed within the surface 10 cm of soil. Higher organic matter levels stabilize soil aggregates, which increases soil tilth. Such benefits do not persist, however, if plowing is occasionally implemented in the conservation tillage system.

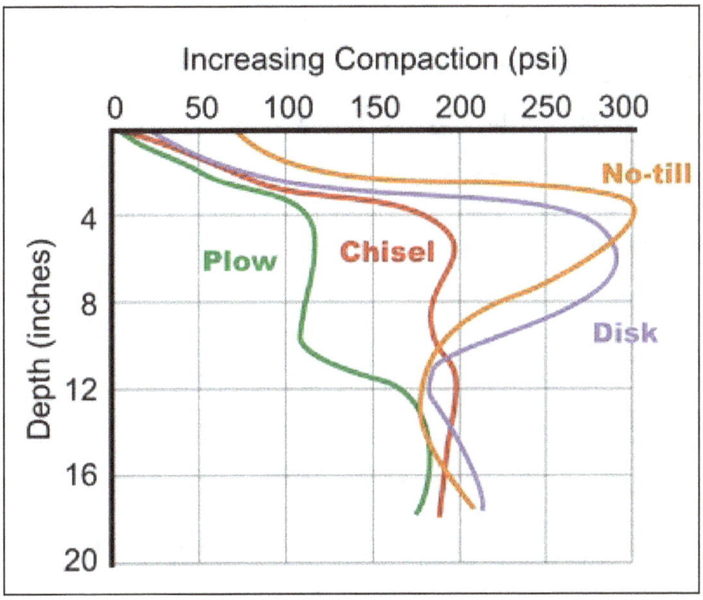

Soil compaction (measured by cone penetrometer) was higher for no-till than for other systems in an Illinois study. Fertilizer management for today's tillage systems

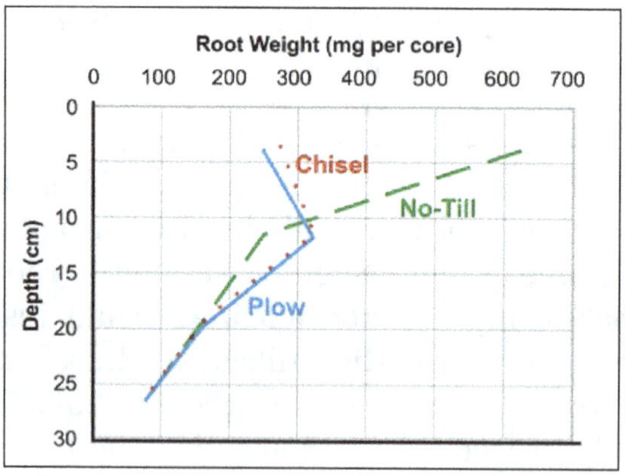

Hoeft, R.G. and G.W. Randall. Tillage affects fertility: How to alter one when you change the other.

- Root Density – The distribution of roots in the soil profile is often dependent upon the tillage system implemented. Roots of plants growing in conservation tillage systems are more concentrated at shallower depths than plant roots grown in other tillage systems. This difference in distribution is attributable in part to the higher soil moisture levels near the surface in conservation tillage. Important also is that without soil mixing, nutrients become more concentrated near the soil surface. Such a nutrient distribution contributes to the concentration of roots nearer the soil surface.

References

- Tillage, practices: soilquality.org, Retrieved 21 April, 2019
- Types-of-tillage, ericeproduction: knowledgebank.irri.org, Retrieved 26 June, 2019
- Typesoftillage, tillage, agriculture: tnau.ac.in, Retrieved 6 July, 2019
- Primary Tillage: hillagric.ac.in, Retrieved 19 January, 2019
- Secondary tillage: hillagric.ac.in, Retrieved 9 March, 2019
- Tillage-meaning, types-objectives-and-concepts-agriculture, crop: yourarticlelibrary.com, Retrieved 1 February, 2019

Chapter 3

Weeds and their Managements

An undesirable plant in a human controlled setting is known as a weed. Its management involves the process of controlling weeds. Some of the different forms of weeds are green foxtail, dodder weed and rag weed. This chapter has been carefully written to provide an easy understanding of the different types of weeds and their management.

Weed

Weed is a general term for any plant growing where it is not wanted. Ever since humans first attempted the cultivation of plants, they have had to fight the invasion by weeds into areas chosen for crops. Some unwanted plants later were found to have virtues not originally suspected and so were removed from the category of weeds and taken under cultivation. Other cultivated plants, when transplanted to new climates, escaped cultivation and became weeds or invasive species. The category of weeds thus is ever changing, and the term is a relative one.

Weeds interfere with a variety of human activities, and many methods have been developed to suppress or eliminate them. These methods vary with the nature of the weed itself, the means at hand for disposal, and the relation of the method to the environment. Usually for financial and ecological reasons, methods used on a golf course or a public park cannot be applied on rangeland or in the forest. Herbicide chemicals sprayed on a roadside to eliminate unsightly weeds that constitute a fire or traffic hazard are not proper for use on cropland. Mulching, which is used to suppress weeds in a home garden, is not feasible on large farms. Weed control, in any event, has become a highly specialized activity. Universities and agricultural colleges teach courses in weed control, and industry provides the necessary technology. In agriculture, weed control is essential for maintaining high levels of crop production.

The many reasons for controlling weeds become more complex with the increasing development of technology. Plants become weeds as a function of time and place. Tall weeds on roadsides presumably were not problematic prior to the invention of the automobile. However, with cars and increasing numbers of drivers on roads, tall weeds became dangerous, potentially obscuring drivers' visibility, particularly at intersections. Sharp-edged grasses are nominal nuisances in a cow pasture; when the area is converted to a golf course or a public park, they become an actual nuisance. Poison oak (Toxicodendron diversilobum) is rather a pleasant shrub on a sunny hillside in the open country; in a camp ground it is a definite health hazard. Such examples could be given ad infinitum to cover every aspect of agriculture, forestry, highway, waterway and public land management, arboretum, park and golf-course care, and home landscape maintenance.

Weeds compete with crop plants for water, light, and nutrients. Weeds of rangelands and pastures may be unpalatable to animals, or even poisonous; they may cause injuries, as with lodging of foxtails (Alopecurus species) in horses' mouths; they may lower values of animal products, as in the cases of cockleburs (Xanthium species) in wool; they may add to the burden of animal care, as when horses graze in sticky tarweeds (Madia species). Many weeds are hosts of plant disease organisms. Examples are prickly lettuce (Lactuca scariola) and sow thistle (Sonchus species) that serve as hosts for downy mildew; wild mustards (Brassica species) that host clubroot of cabbage; and saltbrush (Atriplex species) and Russian thistle, in which curly top virus overwinters, to be carried to sugar beets by leafhoppers. Many weeds are hosts of insect pests, and a number are invasive species.

Characteristics of Weeds

There are approximately 250,000 species of plants worldwide; of those, about 3 percent, or 8,000 species, behave as weeds. Of those 8,000, only 200 to 250 are major problems in worldwide cropping systems. A plant is considered a weed if it has certain characteristics that set it apart from other plant species. Weeds possess one or more of the following characteristics that allow them to survive and increase in nature:

- Abundant seed production.
- Rapid population establishment.
- Seed dormancy.
- Long-term survival of buried seed.
- Adaptation for spread.
- Presence of vegetative reproductive structures.
- Ability to occupy sites disturbed by humans.

Abundant Seed Production

Weeds can produce tens or hundreds of thousands of seeds per plant, while most crop plants only produce several hundred seeds per plant. The following are some examples of approximate numbers of seeds produced per weed:

- Giant foxtail-10,000.

- Common ragweed-15,000.
- Purslane-52,000.
- Lambsquarters-72,000.
- Pigweed-117,000.

Since most weeds deposit their seeds back to the soil, seed numbers in the soil increase rapidly from year to year if the weeds are not managed. Despite that many weed seeds are either not viable, eaten by animals or insects, or decompose within several months after they are deposited, hundreds of millions of viable weed seeds per acre can still be present and waiting to germinate.

Rapid Population Establishment

Most weeds can germinate and become established relatively quickly. They also produce viable seeds even under environmental and soil conditions that are not favorable for most crop plants. Under ideal conditions, dense weed populations can thrive and easily outcompete a crop if left unchecked. Under poor conditions, certain weeds can adapt and produce some viable seeds in a relatively short time period (6 to 8 weeks).

Seed Dormancy

Dormancy is basically a resting stage or a temporary state in which the weed seeds do not germinate because of certain factors. Dormancy is a survival mechanism that prevents germination when conditions for survival are poor. For example, seeds of summer annual weeds will generally not germinate in the fall, preventing them from being killed by cold winter conditions. The various factors that affect dormancy are temperature, moisture, oxygen, light, the presence of chemical inhibitors, tough seed coat, and immature embryos. There are several kinds of dormancy, but the most commonly used terms to describe dormancy are innate, induced, and enforced.

Innate or primary dormancy inhibits germination at the time seeds are shed from the plant. After the seed shatters from the parent plant, time is required for immature embryos to develop, natural inhibitors to leach out, or extremes of temperature to crack hard seed coats and allow germination to occur. These conditions cause innate dormancy, and, once lost, this type of dormancy cannot reoccur.

Induced dormancy is a temporary dormancy that occurs when a seed is exposed to hot or cold temperatures. It continues after temperatures change and prevents germination during the wrong time of year. The dormancy is broken by temperatures opposite of those that induced it.

Summer heat induces dormancy in summer annual weeds such as yellow foxtail and pigweed, preventing germination in the fall. Cold temperatures in fall and winter break this dormancy (usually by mid-winter), and the seeds germinate in spring when conditions are right. In winter annual weeds, the process is reversed.

Dormancy can be induced in many weed seeds when a crop canopy filters sunlight, shading the ground and reducing germination. Dormancy can be induced over and over again for as long as the seeds remain viable.

Enforced dormancy takes place when environmental conditions--cold temperatures, lack of moisture or oxygen, and occasionally a high salt concentration in the soil--are unfavorable. When limitations are removed, seeds germinate freely. Summer annual weed seeds lose their induced dormancy by mid-winter and, if not for the cold temperatures, would germinate at that time.

Seeds of different weed species have various temperature requirements for germination. Common chickweed can germinate under snow cover, while common purslane will not germinate until the soil temperature reaches 70 to 75 °F. Crop seeds are generally planted at or near the optimum soil temperature needed for quick germination--a temperature that is also ideal for some weed seeds.

Seeds require water for germination. Seeds in dry soils may remain dormant even when all other factors promoting germination are favorable.

Oxygen availability also influences a seed's ability to germinate. Water may fill soil pores and exclude air, limiting germination in very wet soils. Soil compaction also may reduce the oxygen supply and prevent seeds from germinating. Deep plowing, tillage, or hoeing can bring buried seeds to the surface, where they readily germinate upon exposure to oxygen.

Long-term Survival of Buried Seed

If conditions are adequate, buried weed seeds have the potential to remain viable for 40 years or more. Broadleaf weed seeds tend to last longer in the soil than grassy weed seed since they usually have tougher seed coats. In most cases, the majority of seeds only exist in the soil for a few years due to germination, decomposition, predator feeding, or other factors. However, with the large number of seeds produced, a small percentage may remain viable for long-term survival.

Adaptation for Spread

Weeds have certain mechanisms for easy dispersal of seeds. Most seeds or seed pods have special structures that allow them to cling, fly, or float. Common cocklebur and burdock seed pods have hooks that attach to animal fur or feathers; curly dock seeds have bladder-like structures that allow them to float; and milkweed, dandelion, and thistle seeds have a feathery pappus that allows them to be carried by the wind. Other weeds, such as jewelweed or snapweed, have pods that "explode" when the seeds are mature, projecting them several feet from the parent plant. Weeds can also be spread when animals or birds eat their fruit and deposit the seeds with their droppings. Weed seeds can be widely spread through crop seeds, grains, feed hay, and straw. These and other human activities probably account for the long-distance spreading of weeds.

Vegetative Reproductive Structures

Most perennial weeds possess special vegetative structures that allow them to reproduce asexually and survive. These perennial structures contain carbohydrates (food reserves, sugars), have numerous buds in which new plants can arise, and include the following:

- Stolons- Aboveground, horizontal stems that root at the nodes (e.g., crabgrass, bermudagrass, ground ivy).
- Rhizomes- Belowground, thickened stems that grow horizontally in the upper soil layers (e.g., quackgrass, Johnsongrass, wirestem muhly, Canada thistle).
- Tubers- Enlarged rhizomes with compressed internodes located at the ends of rhizomes (e.g., yellow nutsedge, Jerusalem artichoke, potato).
- Bulbs- Modified leaf tissues for carbohydrate storage that are located at the base of the stem or below the soil line (e.g., wild garlic, onion).
- Budding roots- Modified roots that can store carbohydrates and grow both vertically and horizontally (e.g., hemp dogbane, Canada thistle).

Despite these vegetative reproductive structures, many perennials also reproduce by seed. Some depend heavily on reproduction by seed (e.g., dandelion), while for others it is less important (e.g., yellow nutsedge).

Ability to Occupy Disturbed Sites

Weeds are very opportunistic. When conditions are adequate, weed seeds germinate and colonize if left unchecked. When a site is disturbed, weeds are usually the first to emerge. If a weed becomes established first, it has the competitive advantage over crop plants or desirable vegetation.

Problems with Weeds

Weeds are troublesome in many ways. Primarily, they reduce crop yield by competing for:

- Water,
- Light,
- Soil nutrients,
- Space,
- CO_2.

The following are other problems associated with weeds:

- Reducing crop quality by contaminating the commodity.
- Interfering with harvest.
- Serving as hosts for crop diseases or providing shelter for insects to overwinter.
- Limiting the choice of crop rotation sequences and cultural practices.
- Producing chemical substances that can be allergins or toxins to humans, animals, or crop plants (allelopathy).
- Producing thorns and woody stems that cause irritations and abrasions to skin, mouths, or hooves of livestock.
- Being unsightly, dominant, aggressive, or unattractive.
- Obstructing visibility along roadways, interfering with delivery of public utilities (power lines, telephone wires), obstructing the flow of water in water ways, and creating fire hazards.
- Accelerating deterioration of recreational areas, parking lots, buildings, and equipment.
- Invading exotic weed species that can displace native species in stabilized natural areas.

Weeds reduce crop yield and quality and compete for necessary resources

Benefits of Weeds

Despite the negative impacts of weeds, some plants usually thought of as weeds may actually provide some benefits, such as:

- Stabilizing and adding organic matter to soils.
- Providing habitat and feed for wildlife.
- Providing nectar for bees.

- Offering aesthetic qualities.
- Serving as a genetic reservoir for improved crops.
- Providing products for human consumption and medicinal use.
- Creating employment opportunities.

Weeds have a controversial nature. But to the agriculturist, they are plants that need to be managed in an economical and practical way in order to produce food, feed, and fiber for humans and animals. In this context, the negative impacts of weeds indirectly affect all living beings.

Weed Ecology and Biology

Only about 40 percent of the weeds found in the United States are native, while the remaining 60 percent are considered exotic or imported.

Origin of Weeds

Weeds are found throughout the world. However, all are not common in every region. Since weeds can be easily spread, more and more are being disseminated to places where they were not originally found. Only about 40 percent of the weeds found in the United States are native, while the remaining 60 percent are considered exotic or imported. The following are some examples of weeds and their origins:

- United States- Common and giant ragweed, common milkweed, fall panicum, common cocklebur, poison ivy, marestail (horseweed), nightshade, wild or common sunflower, and wild onion.
- South America- Pigweed species and prickly sida.
- Europe- Quackgrass, chickweed, Canada thistle, common lambsquarters, common purslane, wild garlic, and yellow foxtail.
- Asia or Africa- Johnsongrass, wild carrot, giant foxtail, velvetleaf, kudzu, and witchweed.

Types of Weeds

Green Foxtail

Green Foxtail is a summer annual that branches at the base to form a summer annual that branches at the base to form a small tuft of leafy culms. In open areas, the slender culms are erect, ascending, or widely spreading; they are 1½–2' long, terete, light green, and hairless. The blades of the alternate leaves are up to 6" long and ½" across; they are light to medium dull green and hairless. Their margins are usually flat and entire, occasionally curving upward and becoming minutely wavy. The open sheaths are light to medium green like the leaves, but slightly more shiny; they wrap around the culms loosely. At the base, the sheaths are often purplish, while their margins are ciliate. At the junction of each blade and sheath, the ligule is defined by a ring of white hairs.

Each culm terminates in a spike-like panicle of florets about 2-4" long and ½" across. The light green panicle has a dense and cylindrical appearance; it is either straight or slightly nodding. The exterior of each panicle is surrounded with ascending bristles that are light green or purple. The lateral branches of the panicle are very short, terminating in individual spikelets. Each fertile spikelet consists of a single floret, a pair of lemmas, and a pair of tiny glumes at the bottom; it is 1.5–2.5 mm. in length (excluding the bristles, which are 3-5 times longer). Each spikelet has 1-3 bristles; these bristles originate from an area directly underneath the spikelet. One lemma extends the full length of the spikelet, while the other lemma is about one-half the length of the spikelet. Some spikelets may fail to produce any florets or grains. The blooming period can occur from late spring to early fall, but lasts only 1-3 weeks for each plant. The florets are wind-pollinated. Each fertile floret produces a single grain about 1.5 mm. long and half as much across. Individual grains are ovoid in shape, flattened on one side, taperedDistribution Map at both ends, and bone-colored to pale brown. The grains fall from the seedhead when they are ripe, while the bristles remain attached. The root system is shallow and fibrous. This grass reproduces by reseeding itself.

- Cultivation: This weedy grass is found in full to partial sun, moist to dry-mesic conditions, and many kinds of soil, including those containing loam, clay, and gravel. After the seeds germinate, this grass develops very quickly. It is can produce several generations of plants in a single growing season.
- Range & Habitat: The non-native Green Foxtail is very common in Illinois; it is found in every county. Habitats include cropland, fallow fields, pastures, gravelly areas along roads and railroads, cracks in urban parking lots and sidewalks, vacant lots, gardens and lawns, barren areas of mined land, and open waste areas. This grass prefers highly disturbed areas where bare ground has been exposed. It does not invade high quality natural areas to any significant extent.

This weedy grass is ubiquitous and can appear just about anywhere. Among the weedy Setaria spp., it is the first one to develop its seedheads and bloom (sometimes as early as late spring). However, later generations of plants will bloom later in the year. Green Foxtail is fairly easy to identify. It is usually a medium-small grass with squiggly light green seedheads; the bristles of young seedheads are either light green or purple (purple bristles seem to be more common during cool weather). These seedheads are fairly straight and there are only 1-3 bristles that originate from underneath each tiny spikelet. Other weedy Setaria spp. have nodding seedheads (Setaria farberi) or seedheads with yellow bristles (Setaria glauca). Furthermore, they often have either 4-12 bristles or only 1 bristle per spikelet. Another common name of Setaria viridis is Green Bristlegrass.

Striga

Striga, commonly known as Witchweed, is a genus of parasitic plants that occur naturally in parts of Africa, Asia, and Australia. It is in the family Orobanchaceae. Some species are serious pathogens of cereal crops, with the greatest effects being in savanna agriculture in Africa. It also causes considerable crop losses in other regions, including other tropical and subtropical crops in its native range and in the Americas.

Witchweeds are characterized by bright-green stems and leaves and small, brightly colored and attractive flowers. They are obligate hemiparasites of roots and require a living host for germination and initial development, though they can then survive on their own. The genus is classified in the family Orobanchaceae, although older classifications place it in the Scrophulariaceae. The number of species is uncertain, but may exceed 40 by some counts.

Hosts and Symptoms

Although most species of Striga are not pathogens that affect human agriculture, some species have devastating effects upon crops, particularly those planted by subsistence farmers. Crops most commonly affected are corn, sorghum, rice and sugarcane. Three species cause the most damage: Striga asiatica, S. gesnerioides, and S. hermonthica.

Witchweed parasitizes maize, millet, sorghum, sugarcane, rice, legumes, and a range of weedy grasses. It is capable of significantly reducing yields, in some cases wiping out the entire crop.

Host plant symptoms, such as stunting, wilting, and chlorosis, are similar to those seen from severe drought damage, nutrient deficiency, and vascular disease.

Lifecycle

Photo of plant roots with connected Striga plant

Each plant is capable of producing between 90,000 and 500,000 seeds, which may remain viable in the soil for over 10 years. Most seeds produced are not viable. An annual plant, witchweed overwinters in the seed stage. Its seeds germinate in the presence of host root exudate, and develop haustoria which penetrate host root cells. Host root exudate contain strigolactones, signaling molecules that promote striga seed germination. A bell-like swelling forms where the parasitic roots attach to the roots of the host. The pathogen develops underground, where it may spend the next four to seven weeks before emergence, when it rapidly flowers and produces seeds. Witchweed seeds spread readily via wind and water, and in soil via animal vectors. The chief means of dispersal, however, is through human activity, by means of machinery, tools, and clothing.

Haustorium Development

Once germination is stimulated, the *Striga* seed sends out an initial root to probe the soil for the host root. The initial root secretes an oxidizing enzyme that digests the host root surface, releasing quinones. If the quinone product is within the appropriate concentrations, a haustorium will develop from the initial root. The haustorium grows toward the host root until it makes contact with the root surface, establishing parasitic contact in relatively short order. Within 12 hours of initial haustorium growth, the haustorium recognizes the host root and begins rapid cell division and elongation. The haustorium forms a wedge shape and uses mechanical force and chemical digestion to penetrate the host root, pushing the host cells out of the way. Within 48–72 hours, the haustorium has penetrated the host root cortex. Finger-like structures on the haustorium, called oscula (from Latin *osculum*, "little mouth") penetrate the host xylem through pits in the membrane. The oscula then swell to secure their position within the xylem membrane. *Striga* sieve tubes develop along with the oscula. Shortly after the host xylem is penetrated, *Striga* sieve tubes develop and approach the host phloem within eight cells. This eight cell layer allows for nonspecific nutrient transport from the host to the *Striga* seedling. Within 24 hours after tapping the host xylem and phloem, the *Striga* cotyledons emerge from the seed.

Environment

Temperatures ranging from 30 to 35 °C (86 to 95 °F) in a moist environment are ideal for germination. Witchweed will not develop in temperatures below 20 °C (68 °F). Agricultural soils with a light texture and low nitrogen levels tend to favor Striga's development. Still, witchweed has demonstrated a wide tolerance for soil types if soil temperatures are favorably high. Seeds have been shown to survive in frozen soil of temperatures as low as −15 °C (5 °F), attesting to their aptitude as overwintering structures.

Soil temperature, air temperature, photoperiod, soil type, and soil nutrient and moisture levels do not greatly deter the development of witchweed. These findings,while limited to the Carolinas in the United States, seem to suggest that the pathogen could successfully infect the massive corn crops of the American Midwest.

Management

Management of witchweed is difficult because the majority of its life cycle takes place below ground. If it is not detected before emergence, it is too late to reduce crop losses.

To prevent witchweed from spreading it is necessary to plant uncontaminated seeds and to clean soil and plant debris off of machinery, shoes, clothing, and tools before entering fields. If populations are low, hand weeding before seeds are produced is an option.

Striga in the United States has been controlled through the use of several management strategies, including quarantines imposed on affected areas, control of movement of farm equipment between infected and uninfected areas, herbicide application, and imposed "suicidal germination". For the latter, in fields not yet planted in crops, seeds present in the soil are induced to germinate by injecting ethylene gas, which mimics the natural physiological response tied to host recognition. Because no host roots are available, the seedlings die. Unfortunately, each *Striga* plant can produce tens of thousands of tiny seeds, which may remain dormant in the soil for many years. Thus, such treatments do not remove all seeds from the soil. Moreover, this method is expensive and not generally available to farmers in developing nations of Africa and Asia.

Another method called trap cropping involves planting a species in an infested field that will induce the *Striga* seeds to germinate but will not support attachment of the parasite. This method has been used in sorghum plantations by planting *Celosia argentea* between the sorghum rows. Cotton, sunflower and linseed are also effective trap crops. Planting silverleaf (*Desmodium uncinatum*) inhibits striga seed germination and has worked effectively intercropped with maize.

Increasing nitrogen levels in the soil, growing striga-tolerant varieties, trap-cropping, and planting susceptible crops harvested before witchweed seed is produced, are proven anti-striga tactics. Coating maize seeds with fungi or a herbicide also appears to be a promising approach. An example is TAN222, the "striga-resistant" maize variety which is coated with the systemic herbicide imazapyr, to which it is resistant. Any striga seeds sprouting when this maize is in the seedling stage are poisoned when their haustoria embed in the seedling's roots.

Several sorghum varieties have high levels of resistance in local conditions, including 'N-13', 'Framida', and 'Serena'. 'Buruma', 'Shibe', 'Okoa' and 'Serere 17' millet cultivars are considered to be resistant in Tanzania. Some maize varieties show partial resistance to witchweed, including 'Katumani' in Kenya. In a number of rice cultivars, including some cultivars of NERICA (New Rice for Africa), effective pre- and post- attachment resistance mechanisms have been identified. 'StrigAway'™ herbicide-resistant, herbicide impregnated maize has been shown to reduce the seed bank of striga by 30% in two seasons.

Importance

Maize, sorghum, and sugarcane crops affected by witchweed in the United States have an estimated value well over $20 billion. Furthermore, witchweed is capable of wiping out an entire crop. In fact, it is so prolific that in 1957 Congress allocated money in an attempt to eradicate witchweed. Because of Striga, the Animal and Plant Health Inspection Service (APHIS) of the U.S. Department of Agriculture established a research station and control methods. Through infestation mapping, quarantine, and control activities such as contaminated seed destruction, the acreage parasitized by witchweed has been reduced 99% since its discovery in the United States. APHIS has even offered cash rewards those who identify and report the weed, and encourages landowners to check their own acreage.

Parasitizing important economic plants, witchweed is one of the most destructive pathogens in Africa. In fact, witchweed affects 40% of Africa's arable savanna region, resulting in up to $13 billion lost every year. Striga affects 40 million hectares (98,842,153 acres) of crops in sub-Saharan Africa alone. The witchweed infestation is so bad in parts of Africa, some farmers must relocate every few years. The majority of crops in Africa are grown by subsistence farmers who cannot afford expensive witchweed controls, who therefore suffer much as a result of this pathogen.

Species

Common Crop Parasites

Photo of a Striga asiatica plant (witchweed)

Striga bilabiata

- *Striga asiatica* has a very wide geographic distribution, from Africa through southern and eastern Asia to Australia. Since the 1950s, it is also known from the United States. This introduction, likely a result of human activity, resulted in an infestation of corn (maize) across many counties in North and South Carolina. The United States Department of Agriculture and state agencies imposed a quarantine on this area to control its spread - a process that was apparently successful.
- *Striga gesnerioides*, cowpea witchweed, as its name implies, is a parasite of cowpea (*Vigna unguiculata*), which is not a grass, but a member of the legume family (Fabaceae or Leguminosae). This species was also accidentally introduced into Florida in the United States, where it was found parasitizing *Indigofera hirsuta* (hairy indigo, another legume).
- *Striga hermonthica* (purple witchweed) is also a parasite that affects grasses, particularly sorghum and pearl millet in sub-Saharan Africa (Senegal to Ethiopia, Democratic Republic of Congo and Tanzania, Angola, Namibia).

Orange Hawkweed

Orange hawkweed is a perennial with a shallow root system. Leaves are hairy, up to 6 inches long, and found at the base of the flower. Each rosette is capable of producing between 10 and 30 flowering stems, and each stem is capable of producing 5 to 30 flowering heads. A single flower head is capable of producing between 12 and 50 tiny black seeds. Orange hawkweed has a shallow root system and underground creeping stems called rhizomes. New plants can arise from buds on the rhizomes and plants can develop several creeping stems (like strawberries) that are also capable of producing new plants.

Orange Hawkweed flowers.

Orange Hawkweed rosettes.

Hawkweeds spread by both runners and seed, with a one square metre mat producing up to 40,000 seeds a year. In New Zealand H. pilosella, which is similar in habit and form to orange hawkweed, spreads mainly by runners rather than seed. Runners begin to grow from buds in the rosette leaves when the plants are flowering. Depending on growing conditions, plants produce between four and eight leafy runners that can reach lengths of 100–250 mm. These runners form new rosettes and, once established, the patch continues to expand until it covers the site with a solid mat of rosettes. Under ideal conditions, it can form a colony up to 0.5 m across in its first year. Hawkweeds regrow each year from short, below-ground rhizomes.

Seeding is important for more widespread dispersal. Minute barbs along ribs on the seeds enable them to stick to hair, fur, feathers, clothing and vehicles, and be carried long distances. Seeds can be dispersed by wind and water, and in dumped garden waste and contaminated soil.

Dodder Weed

Dodder weed has thin, twining stems either pale green, yellow or bright orange in color. It is either leaess or with tiny, triangular leaves. The weed bares cream colored bell-shaped blooms containing a seed capsule with 2-3 seeds.

The rootless seedlings have a limited ability to photosynthesize and rely on the host plant to provide them with energy. A dodder seedling can live 5-10 days without a host, but will soon die. As the dodder weed grows, it continually reattaches itself to its host and sends out shoots to attach to nearby hosts as well creating a dense mass of intertwined stems.

Seeds are generally dispersed via the movement of soil and equipment or in dirt clinging to shoes and tires, or in infested plant material that is being transported. The seed germinates at or near the soil surface in the spring when temps reach 60 degrees F. (15 C). Upon emergence, the seedling is dependent upon carbohydrates stored in the seed until they attach to a host. Once attached, dodder plant extracts nutrients and water from the host, predisposing the host to disease and insect invasion, aecting fruit set and yield and ev aecting fruit set and yield and even killing o the host.

Dodder Weed Control

Dodder is a difficult weed to eradicate. It grows rapidly and can set seed after only a few weeks of growth. Its seed can survive in the soil for long periods. Preventing it from entering your property and being vigilant in spotting and destroying new outbreaks before they seed is the best form of control. A long-term approach is required to manage this weed. Several options are available, depending on the size of the infestation, the crop or pasture infested, and the crop or pasture rotation available. Whichever option is chosen, it is essential that treatment should begin as soon as possible after an infestation is noticed. Scattered plants can rapidly result in a major infestation which is difficult and costly to control.

Prevention

Stopping the entry of dodder onto your farm is the best control measure. Early detection is necessary if dodder is to be controlled before seeding commences. Lucerne paddocks should be inspected thoroughly before first cut, to detect dodder growth. Pasture paddocks need to be inspected after rainfall in late spring. If dodder is found, control strategies need to be implemented immediately. Mapping existing infestations at the end of summer will help in checking for reinfestations the following spring.

Sow Clean Seed

Pasture seed infested with dodder is an important source of new infestations. Because of their similar size, dodder seed is very difficult to remove from lucerne and small clover seeds. It is therefore essential to use certified pasture seed.

Buy Clean Fodder

Contaminated hay and grain can introduce dodder to a clean property. Hay containing dodder should be destroyed as mature seed is often present.

Restrict Movement of Seed and Plant Fragments

Machinery used for fodder conservation in infested paddocks should be cleaned before moving to other paddocks. Stock grazing on areas known to be infested with dodder should be kept in quarantine for at least two days before they are moved to dodder-free pastures. Wherever infestations are found they should be marked and the area quarantined from activities such as hay-cutting and grazing. Care should also be taken to avoid spreading seed in soil, or on machinery or boots.

Restrict Movement in Irrigation

Small dodder seed can be picked up in irrigation suction lines. Self-cleaning or flushing filters can be used if there is a risk of spreading dodder seed in irrigation water.

Control Weed Hosts

Wireweed, Noogoora burr and Bathurst burr are preferred weed hosts for dodder. Controlling these will help to reduce the risk of new infestations and the extent of current infestations.

Buffer Zones

Since dodder can be spread by seed and stem fragments, it is important to always treat a buffer zone outside the obvious areas of infestation. Look for small threads starting to invade further into the crop. Even small threads up to 5 cm can re-attach to the host crop and continued infestation will result. A buffer zone of at least 1 m from the infestation should be treated with herbicide or burnt. Once treated, infestations should remain marked so the effectiveness of the control can be checked at a later date.

Cutting and Burning

Slashing promotes the spread of dodder. Do not cut or slash dodder and leave the cut material in the crop as it will re-attach itself to a new host. All host and dodder plant material must be burnt, preferably on the infested site. Cut the host plant as close as possible to ground level and burn it. Burning can be carried out after cutting and drying, using flame throwers or other combustible material to ignite the vegetation.

Burning is more effective following the application of a desiccant herbicide. The area should be inspected again after one week, as dodder may re-shoot from the crowns of lucerne and weeds such as wireweed, Noogoora burr or Bathurst burr.

Hotter burns will achieve higher dodder kill rates but have greater risk of killing host crops. A lucerne stand that has been burnt can regenerate from undamaged crowns without the need for re-sowing.

Crop Rotation

Lucerne or clover stands that are severely infested should be cultivated and replaced with less susceptible crops or pastures.

Cereal crops such as wheat, barley, oats, triticale and cereal rye are less susceptible to dodder, along with summer grain crops such as maize and sorghum.

While dodder cannot complete its lifecycle on cereal crops, it may do so on the broadleaf weeds that invade these crops. The advantage of a crop rotation for dodder control is lost if host weeds occur in the crop. In-crop weed control then becomes an essential part of crop management.

If winter cereals are to be grown in an infested paddock a clean fallow must be maintained over the summer months when dodder is active. This can be achieved with herbicides or cultivation.

Pasture rotations will need to be based on perennial grass species where practical. Subterranean clover can be used as the legume component, but any summer-growing broadleaf weed or legume will need to be controlled to prevent dodder reinvading.

Where crops or pastures are used in rotations to control dodder, lucerne or susceptible legumes should not be sown for at least five years.

Deep ploughing can help reduce the seed burden by burying dodder seed. Most dodder seed will not germinate from a depth of more than 7.5 cm.

Infestations of dodder are less likely to occur during the establishment period if lucerne or susceptible clovers are re-sown in late summer to autumn.

Follow-up Control

It is essential that a follow-up control program be implemented even after control measures have been carried out. Lucerne growers in particular should be wary of reinfestation. Under ideal conditions, the seed of dodder can survive in the soil for a number of years after a successful control program.

Ragweed

Ragweeds are flowering plants in the genus *Ambrosia* in the aster family, Asteraceae. They are distributed in the tropical and subtropical regions of the Americas, especially North America, where the origin and center of diversity of the genus are in the southwestern United States and northwestern Mexico. Several species have been introduced to the Old World and some have naturalized and have become invasive species. Ragweed species are expected to continue spreading across Europe in the near future in response to ongoing climate change.

Other common names include bursages and burrobrushes. The genus name is from the Greek *ambrosia*, meaning "food or drink of immortality".

Ragweed pollen is notorious for causing allergic reactions in humans, specifically allergic rhinitis. Up to half of all cases of pollen-related allergic rhinitis in North America are caused by ragweeds. The most widespread species of the genus in North America is *Ambrosia artemisiifolia*.

Description and Ecology

Ragweeds are annual and perennial herbs and shrubs. Species may grow just a few centimeters tall or well exceed four meters in height. The stems are erect, decumbent or prostrate, and many grow from rhizomes. The leaves may be arranged alternately, oppositely, or both. The leaf blades come in many shapes, sometimes divided pinnately or palmately into lobes. The edges are smooth or toothed. Some are hairy, and most are glandular.

Ragweeds are monoecious, most producing inflorescences that contain both staminate and pistillate flowers. Inflorescences are often in the form of a spike or raceme made up mostly of staminate flowers with some pistillate clusters around the base. Staminate flower heads have stamens surrounded by whitish or purplish florets. Pistillate flower heads have fruit-yielding ovules surrounded by many phyllaries and fewer, smaller florets. The pistillate flowers are wind pollinated, and the fruits develop. They are burs, sometimes adorned with knobs, wings, or spines.

Many *Ambrosia* species occur in desert and semi-desert areas, and many are ruderal species that grow in disturbed habitat types.

Allergy

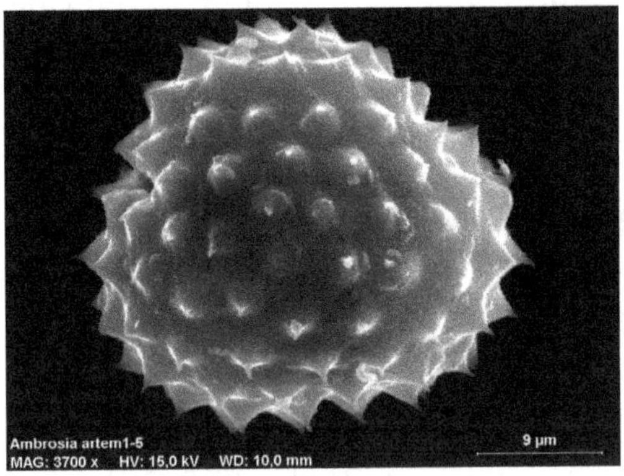

Ambrosia artemisiifolia pollen

Ragweed pollen is a common allergen. A single plant may produce about a billion grains of pollen per season, and the pollen is transported on the wind. It causes about half of all cases of pollen-associated allergic rhinitis in North America, where ragweeds are most abundant and diverse. Common culprits are common ragweed (*A. artemisiifolia*) and great ragweed (*A. trifida*).

Concentration of ragweed pollen—in the absence of significant rainfall, which removes pollen from the air—is the lowest in the early morning hours (6:00 AM), when emissions starts, and pollen concentration peaks at midday. Ragweed pollen can remain airborne for days and travel great distances, affecting people hundreds of miles away. It can even be carried 300 to 400 miles (640 km) out to sea. Ragweeds native to the Americas have been introduced to Europe starting in the nineteenth century and especially during World War I, and have spread rapidly there since the 1950s. Eastern Europe, particularly Hungary, has been badly affected by ragweed since the early 1990s, when the dismantling of Communist collective agriculture led to large-scale abandonment of agricultural land, and new building projects also resulted in disturbed, un-landscaped acreage.

The major allergenic compound in the pollen has been identified as *Amb a 1*, a 38 kDa nonglycosylated protein composed of two subunits. It also contains other allergenic components, such as profilin and calcium-binding proteins.

Ragweed allergy sufferers may show signs of oral allergy syndrome, a food allergy classified by a cluster of allergic reactions in the mouth in response to the consumption certain fruits, vegetables, and nuts. Foods commonly involved include beans, celery, cumin, hazelnuts, kiwifruit, parsley, potatoes, bananas, melons, cucumbers, and zucchini. Because cooking usually denatures the proteins that cause the reaction, the foods are more allergenic when eaten raw; exceptions are celery and nuts, which may not be safe even when cooked. Signs of reaction can include itching, burning, and swelling of the mouth and throat, runny eyes and nose, hives, and, less commonly, vomiting, diarrhea, asthma, and anaphylaxis. These symptoms are due to the abnormal increase of IgE antibodies which attach to a type of immune cell called mast cells. When the ragweed antigen then attaches to these antibodies the mast cells release histamine and other symptom evoking chemicals.

Merck & Co, under license from allergy immunotherapy (AIT) company ALK, has launched a ragweed allergy immunotherapy treatment in sublingual tablet form in the US and Canada. Allergy immunotherapy treatment involves administering doses of the allergen to accustom the body to induce specific long-term tolerance.

Control and Eradication

Chemical spraying has been used for control in large areas. Because ragweed only reacts to some of the more aggressive herbicides, it is highly recommended to consult professionals when deciding on dosage and methodology, especially near urban areas. Effective active ingredients include those that are glyphosate-based (Roundup, Glyphogan, Glialka), sulfosate-based (Medallon), and glufosinate ammonium-based (Finale 14SL). In badly infested areas, 2 to 6.5 liters per hectare (0.2–0.7 U.S. gal/acre) are usually dispersed. In 2007 several *Ambrosia artemisiifolia* populations were glyphosate resistant, exclusively in the USA.

Where herbicides cannot be used, mowing may be repeated about every three weeks, as it grows back rapidly. In the past, ragweed was usually cut down, left to dry, and then burned. This method is used less often now, because of the pollution caused by smoke. Manually uprooting ragweed is generally ineffective, and skin contact can cause allergic reaction. If uprooting is the method of choice, it should be performed before flowering. There is evidence that mechanical and chemical control methods are actually no more effective in the long run than leaving the weed in place.

Fungal rusts and the leaf-eating beetle *Ophraella communa* have been proposed as agents of biological pest control of ragweeds, but the latter may also attack sunflowers, and applications for permits and funding to test these controls have been unsuccessful. The beetle has, however, appeared in Europe, either on its own or as an uncontrolled introduction, and it has started making a dent into Ambrosia populations there.

Deer Tongue

Deer-tongue grass is a wide-leaved, clump-forming perennial grass, characterized by stiff leaves and long, bristly leaf sheaths, some of which pull away from the stem at a characteristic angle. This species reproduces by seed, and often forms large clumps.

Root System

The root system is composed of thickened root stalks.

Stems

Stems are erect, hairy, round and covered with minute bumps. The stems often branch from the upper joints (nodes), and grow 1 1/2 to 3 1/2 feet tall (sometimes up to 5 feet).

Leaves

The leaf blade (free part of the leaf) is stiff, strongly veined, 4 to 8 inches long (sometimes up to 12 inches) and broad (3/4 to 1 1/4 inches wide), tapering to a point. Both upper and lower leaf surfaces are smooth, but the leaf margins are rough. The base of the blade (where it meets the stem) is heart-shaped. Leaf sheaths (part of the leaf surrounding the stem) are round, and those of the lower and middle leaves are long and covered with short, bristle-like hairs. Sheaths often pull away from the stem at a characteristic 45-degree angle to 90-degree angle. The ligule (projection inside on the top of the sheath) is short (1/32 inch long) and membranous, with a fringe of minute hairs. Auricles (appendages at the top of the sheath) are absent.

Flowers

Small, oval flowers are borne on fine stalks in an open, sparse, branched flower head, 3 to 6 inches long. Flower heads that are produced in late summer and fall are often almost completely enclosed within the leaf sheaths.

Fruits and Seeds

Seeds are smooth and 1/10 inch long.

Biology

Flowering occurs from May to September. Deer-tongue grass often forms large patches of stiff foliage that persist through winter.

Bull Thistle

Bull Thistle is an obligatory biennial that forms a rosette of leaves during the first year, and bolts upward during the second year to produce flowers. It forms occasional side stems, but remains erect in stature, reaching 3-6' in height. The stout stems are light green, somewhat angular, and covered with dense white hairs. The alternate leaves are up to 7" long and 2" across. In outline, they are lanceolate in shape, but deeply pinnatifid. The widely spaced lobes narrow into points that are individually armed with a pale yellow spine. Some of the upper leaves near the flowerheads may be lanceolate or linear, but remain unlobed. Across the surface of the leaves, there are short white hairs and scattered small spines. The upper surface is dark green, while the lower surface is light green. At the base of each leaf, there are a pair of narrow wings that are dark green and decurrent against the stem. These wings are extensions of the leaves and arm the stems with spines up to 1/4" long.

The upper stems terminate in flowerheads of purplish pink to purple flowers. Each flowerhead is 1½–2" across and contains numerous disk florets, but no ray florets. Each disk floret is long and tubular, but becomes divided into 5 long thread-like lobes. This provides the flowerhead with a showy hair-like appearance. At the base of each flowerhead, are numerous green bracts that taper into stiff points that curl outward and are quite prickly. Among these bracts are cobwebby hairs in the background. The blooming period occurs from mid- to late summer and lasts about 3-4 weeks, after which the florets wither away. They are replaced by achenes with large tufts of white hair. These achenes are long, slender, and slightly ribbed. The Distribution Maps are dispersed by the wind. The root system consists of a stout taproot that runs deep into the ground. This plants spreads by reseeding itself and occasionally forms colonies.

- Cultivation: This thistle usually grows in full sun under moist to dry conditions. It prefers a fertile soil that consists of loam, clay loam, or that is slightly stony. Occasionally the foliage is affected by mildew during the summer or fall. Individual plants can produce a great abundance of seed, which have a high germination and survival rate. Thus, this plant can be quite aggressive. Because the seeds remain viable for only 1-2 years, one control strategy consists of destroying individual plants before they reach the flowering stage. It is possible for a plant to reestablish itself if a portion of the taproot remains in the ground.

Identification

- Branching, erect biennial, 2 to 6 feet tall.
- Bull thistle flower - click for larger image.
- Rosettes form in first year, flowering stems the second.
- Long, sharp spines on the leaves at the midrib and the tips of the lobes.
- Leaves are deeply lobed and hairy - there are coarse hairs on leaf tops, making leaf feel rough to the touch, and woolly hairs on the underside.
- Leaf bases extend down onto stems and form spiny wings along the stems.
- Pink-magenta flower heads top each stem.
- Flower heads are "gumdrop" shaped and spines extend all around the base of the flower heads.
- Flowering June to September.

Habitat and Impact

Bull thistle prefers sunny, open areas and can tolerate a wide range of conditions, from moist to dry soils, and is typically found in disturbed areas such as roadsides, trails, logged areas, vacant land, pastures and cultivated land. Overgrazed pastures are susceptible to bull thistle encroachment, and it can sometimes form dense stands that reduce productivity and stocking levels. Bull thistle may also dominate forest clear cuts and reduce growth of tree seedlings.

Growth and Reproduction

Bull thistle has a two year life cycle, flowering and setting seed in the second year. Seeds are short-lived on the soil surface but can persist for many years when they are buried, such as form cultivation activities. Seed germination generally occurs in the fall and spring. Basal rosettes form and continue to grow until winter and can grow quite large, up to 3 feet in diameter. Rosettes that are not large enough by spring may not flower until the following year. Flowering usually starts in mid-June and continues into early fall. Plants can be self-pollinated or insect-pollinated. Bull thistle does not reproduce vegetatively and does not have rhizomes.

Control

1. Prevention: Bull thistle only reproduces by seed so prevention of seeding and takign care not to spread seeds are key to preventing new infestations. Contaminated hay is a primary means of spread of this species so be careful to purchase weed free hay or watch closely for new plants in the areas hay is kept or spread. Do not leave cut stems of flowering bull thistle on the ground because they are likely to form viable seed after they are cut.

2. Manual control: Bull thistle can be dug up with a shovel. Usually removing the top couple of inches of root is sufficient to kill the plant, especially after it has bolted (produced stems). A shovel or other tool can used to chop off leaves from one side of the plant to gain easier access to the roots, which can then be dug up. Flowering stems should be collected and destroyed to keep them from forming viable seed.

3. Mechanical control: Mowed thistles will produce new branches from basal buds but close cutting or cutting twice per season will usually prevent seed production and reduce the population over time. More more effective control, cut plants with a sharp shovel at 1-2 inches below the soil surface prior to flowering. If only one cutting a year is possible, cut when plants are in bud for best results. Cultivation and tilling can also be effective to control bull thistle.

4. Biological control: The bull thistle seed head gall fly (Urophora stylata) lays eggs on closed flower buds in June and July. After hatching, the larvae burrow into the seed-prodcuing tissues to feed, forming galls and reducing seed production. If the bull thistle population is large enough to support a good sized population of this insect, it can be an effective way to reduce seed production of the bull thistle. This insect will not get rid of the bull thistle, however, just reduce its impact.

5. Cultural/grazing: Bull thistle is not as competitive in a well-managed pasture and regular cutting combined with good pasture management can reduce bull thistle to a manageable level. Also, goats and sheep have both been used for grazing management of bull thistle. Even horses will help by picking out many of the nectar-rish flowers and eating them before they go to seed.

Ailanthus Altissima

Ailanthus altissima, commonly known as tree of heaven, ailanthus, varnish tree, or in Chinese as chouchun is a deciduous tree in the family Simaroubaceae. It is native to both northeast and central China, as well as Taiwan. Unlike other members of the genus *Ailanthus*, it is found in temperate climats rather than the tropics. The tree grows rapidly and is capable of reaching heights of 15 metres (49 ft) in 25 years. While the species rarely live more than 50 years, some specimens exceed 100 years old. Its remarkable suckering ability makes it possible for this tree to clone itself almost indefinitely. It is considered a noxious weed and vigorous invasive species.

In China, the tree of heaven has a long and rich history. It was mentioned in the oldest extant Chinese dictionary and listed in countless Chinese medical texts for its purported ability to cure various ailments. The roots, leaves and bark are used in traditional Chinese medicine, primarily as an astringent. The tree has been grown extensively both in China and abroad as a host plant for the ailanthus silkmoth, a moth involved in silk production. Ailanthus has become a part of western culture as well, with the tree serving as the central metaphor.

The tree was first brought from China to Europe in the 1740s and to the United States in 1784. It was one of the first trees brought west during a time when *chinoiserie* was dominating European arts, and was initially hailed as a beautiful garden specimen. However, enthusiasm soon waned after gardeners became familiar with its suckering habits and its foul smelling odor. Despite this, it was used extensively as a street tree during much of the 19th century. Outside Europe and the United States the plant has been spread to many other areas beyond its native range. In a number of these, it has become an invasive species due to its ability both to colonize disturbed areas quickly and to suppress competition with allelopathic chemicals. It is considered a noxious weed in Australia, the United States, New Zealand and many countries of central, eastern and southern Europe. The tree also resprouts vigorously when cut, making its eradication difficult and time-consuming. In many urban areas, it has acquired the derisive nicknames of "ghetto palm", "stink tree", and "tree of Hell".

A. altissima is a medium-sized tree that reaches heights between 17 and 27 metres (56 and 89 ft) with a diameter at breast height of about 1 metre (39 inches). The bark is smooth and light grey, often becoming somewhat rougher with light tan fissures as the tree ages. The twigs are stout, smooth to lightly pubescent, and reddish or chestnut in color. They have lenticels as well as heart-shaped leaf scars (i.e. a scar left on the twig after a leaf falls) with many bundle scars (i.e. small marks where the veins of the leaf once connected to the tree) around the edges. The buds are finely pubescent, dome shaped, and partially hidden behind the petiole, though they are completely visible in the dormant season at the sinuses of the leaf scars. The branches are light to dark gray in color, smooth, lustrous, and containing raised lenticels that become fissures with age. The ends of the branches become pendulous. All parts of the plant have a distinguishing strong odor that is often likened to peanuts, cashews, or rotting cashews.

The leaves are large, odd- or even-pinnately compound, and arranged alternately on the stem. They range in size from 30 to 90 cm (0.98 to 2.95 ft) in length and contain 10–41 leaflets organised in pairs, with the largest leaves found on vigorous young sprouts. When they emerge in the spring, the leaves are bronze then quickly turn from medium to dark green as they grow. The rachis is light to reddish-green with a swollen base. The leaflets are ovate-lanceolate with entire margins, somewhat asymmetric and occasionally not directly opposite to each other. Each leaflet is 5 to 18 cm (2.0 to 7.1 in) long and 2.5 to 5 cm (0.98 to 1.97 in) wide. They have a long tapering end while the bases have two to four teeth, each containing one or more glands at the tip. The leaflets' upper sides are dark green in color with light green veins, while the undersides are a more whitish green. The petioles are 5 to 12 mm (0.20 to 0.47 in) long. The lobed bases and glands distinguish it from similar sumac species.

Bark and flowers of *A. altissima*

Immature seeds on a female tree.

Tree of Heaven Re-sprouting

The flowers are small and appear in large panicles up to 50 cm (20 in) in length at the end of new shoots. The individual flowers are yellowish green to reddish in color, each with five petals and sepals. The sepals are cup-shaped, lobed and united while the petals are valvate (i.e. they meet at the edges without overlapping), white and hairy towards the inside. They appear from mid-April in the south of its range to July in the north. *A. altissima* is dioecious, with male and female flowers being borne on different individuals. Male trees produce three to four times as many flowers as the females, making the male flowers more conspicuous. Furthermore, the male plants emit a foul-smelling odor while flowering to attract pollinating insects. Female flowers contain ten (or rarely five through abortion) sterile stamens (stamenoides) with heart-shaped anthers. The pistil is made up of five free carpels (i.e. they are not fused), each containing a single ovule. Their styles are united and slender with star-shaped stigmas.

The male flowers are similar in appearance, but they of course lack a pistil and the stamens do function, each being topped with a globular anther and a glandular green disc. The fruits grow in clusters; a fruit cluster may contain hundreds of seeds. The seeds borne on the female trees are 5 mm in diameter and each is encapsulated in a samara that is 2.5 cm (0.98 in) long and 1 cm (0.39 in) broad, appearing July though August, but can persist on the tree until the next spring. The samara is large and twisted at the tips, making it spin as it falls, assisting wind dispersal, and aiding buoyancy for long-distance dispersal through hydrochory. Primary wind dispersal and secondary water dispersal are usually positively correlated in *A. altissima* since most morphological characteristics of samaras affect both dispersal modes in the same way – except for the width of the samaras, which in contrast affects both types of dispersal in opposing ways, allowing differentiation in the dispersal strategies of this tree. The females can produce huge amounts of seeds, normally around 30,000 per kilogram (14,000/lb) of tree, and fecundity can be estimated non-destructively through measurements of dbh.

Distribution and Habitat

A. altissima is native to northern and central China, Taiwan and northern Korea. It was historically widely distributed, and the fossil record indicates clearly that it was present in North America as recently as the middle Miocene. In Taiwan it is present as var. *takanai*. In China it is native to every province except Gansu, Heilongjiang, Hainan, Jilin, Ningxia, Qinghai, Xinjiang, and Tibet. It has been introduced in many regions across the world.

The tree prefers moist and loamy soils, but is adaptable to a very wide range of soil conditions and pH values. It is drought-hardy, but not tolerant of flooding. It also does not tolerate deep shade. In China it is often found in limestone-rich areas. The tree of heaven is found within a wide range of climatic conditions. In its native range it is found at high altitudes in Taiwan as well as lower ones in mainland China. These are virtually found anywhere in the U.S., but especially in arid regions bordering the Great Plains, very wet regions in the southern Appalachians, cold areas of the lower Rocky Mountains and throughout much of the California Central Valley, forming dense thickets that displace native plants. Prolonged cold and snow cover cause dieback, although the trees resprout from the roots.

Ailanthus altissima growing in Australia.

An Exotic Plant

The earliest introductions of *A. altissima* to countries outside of its native range were to the southern areas of Korea as well as to Japan. It is possible that the tree is native to these areas, but it is generally agreed that the tree was a very early introduction. Within China itself it has also been naturalized beyond its native range in areas such as Qinghai, Ningxia and Xinjiang.

In 1784, not long after Jussieu had sent seeds to England, some were forwarded to the United States by William Hamilton, a gardener in Philadelphia. In both Europe and America it quickly became a favoured ornamental, especially as a street tree, and by 1840 it was available in most nurseries. The tree was separately brought to California in the 1890s by Chinese immigrants who came during the California Gold Rush. It has escaped cultivation in all areas where it was introduced, but most extensively in the United States. It has naturalized across much of Europe, including Germany, Austria, Switzerland, the Czech Republic, the Pannonian region (i.e. southeastern Central Europe around the Danube river basin from Austria, Slovakia and Hungary south to the Balkan ranges) and most countries of the Mediterranean Basin. In Montenegro and Albania *A. altissima* is widespread in both rural and urban areas and while in the first it was introduced as an ornamental plant, it very soon invaded native ecosystems with disastrous results and became an invasive species. Ailanthus has also been introduced to Argentina, Australia (where it is a declared weed in New South Wales and Victoria), New Zealand (where it is listed under the National Pest Plant Accord and is classed an "unwanted organism"), the Middle East and in some countries in South Asia such as Pakistan. In South Africa it is listed as an invasive species which must be controlled, or removed and destroyed.

In North America, *A. altissima* is present from Massachusetts in the east, west to southern Ontario, southwest to Iowa, south to Texas, and east to the north of Florida. On the west coast it is found from New Mexico west to California and north to Washington. In the east of its range it grows most extensively in disturbed areas of cities, where it was long ago present as a planted street tree. It also grows along roads and railways. For example, a 2003 study in North Carolina found the tree of heaven was present on 1.7% of all highway and railroad edges in the state and had been expanding its range at the rate of 4.76% counties per year. Similarly, another study conducted in southwestern Virginia determined that the tree of heaven is thriving along approximately 30% of the state's interstate highway system length or mileage. It sometimes enters undisturbed areas as well and competes with native plants. In western North America it is most common in mountainous areas around old dwellings and abandoned mining operations. It is classified as a noxious or invasive plant on National Forest System lands and in many states because its prolific seed production, high germination rates and capacity to regrow from roots and root fragments enable *A. altissima* to out-compete native species. For this reason, control measures on public lands and private property are advised where *A. altissima* has naturalized.

Ecology

A female bearing a heavy load of seeds in Valladolid, Spain

Tree of heaven is an opportunistic plant that thrives in full sun and disturbed areas. It spreads aggressively both by seeds and vegetatively by root sprouts, re-sprouting rapidly after being cut. It is considered a shade-intolerant tree and cannot compete in low-light situations, though it is sometimes found competing with hardwoods, but such competition rather indicates it was present at the time the stand was established. On the other hand, a study in an old-growth hemlock-hardwood forest in New York found that Ailanthus was capable of competing successfully with native trees in canopy gaps where only 2 to 15% of full sun was available. The same study characterised the tree as using a "gap-obligate" strategy in order to reach the forest canopy, meaning it grows rapidly during a very short period rather than growing slowly over a long period. It is a short lived tree in any location and rarely lives more than 50 years. Ailanthus is among the most pollution-tolerant of tree species, including sulfur dioxide, which it absorbs in its leaves. It can withstand cement dust and fumes from coal tar operations, as well as resist ozone exposure relatively well. Furthermore, high concentrations of mercury have been found built up in tissues of the plant.

Ailanthus has been used to re-vegetate areas where acid mine drainage has occurred and it has been shown to tolerate pH levels as low as 4.1 (approximately that of tomato juice). It can withstand very low phosphorus levels and high salinity levels. The drought-tolerance of the tree is strong due to its ability to effectively store water in its root system. It is frequently found in areas where few trees can survive. The roots are also aggressive enough to cause damage to subterranean sewers and pipes. Along highways it often forms dense thickets in which few other tree species are present, largely due to the toxins it produces to prevent competition.

Female tree growing in Chicago, Illinois

Ailanthus produces an allelopathic chemical called ailanthone, which inhibits the growth of other plants. The inhibitors are strongest in the bark and roots, but are also present in the leaves, wood and seeds of the plant. One study showed that a crude extract of the root bark inhibited 50% of a sample of garden cress (*Lepidium sativum*) seeds from germinating. The same study tested the extract as an herbicide on garden cress, redroot pigweed (*Amaranthus retroflexus*), velvetleaf (*Abutilon theophrasti*), yellow bristlegrass (*Setaria pumila*), barnyard grass (*Echinochloa crusgalli*), pea (*Pisum sativum* cv. Sugar Snap) and maize (*Zea mays* cv. Silver Queen). It proved able to kill nearly 100% of seedlings with the exception of velvetleaf, which showed some resistance. Another experiment showed a water extract of the chemical was either lethal or highly damaging to 11 North American hardwoods and 34 conifers, with the white ash (*Fraxinus americana*) being the only plant not adversely affected. The chemical does not, however, affect the tree of heaven's own seedlings, indicating that *A. altissima* has a defence mechanism to prevent autotoxicity. Resistance in various plant species has been shown to increase with exposure. Populations without prior exposure to the chemicals are most susceptible to them. Seeds produced from exposed plants have also been shown to be more resistant than their unexposed counterparts.

The tree of heaven is a very rapidly growing tree, possibly the fastest growing tree in North America. Growth of one to two metres (3.3 to 6.6 ft) per year for the first four years is considered normal. Shade considerably hampers growth rates. Older trees, while growing much slower, still do so faster than other trees. Studies found that Californian trees grew faster than their East Coast counterparts, and American trees in general grew faster than Chinese ones.

In northern Europe the tree of heaven was not considered naturalised in cities until after the Second World War. This has been attributed to the tree's ability to colonize areas of rubble of destroyed buildings where most other plants would not grow. In addition, the warmer microclimate in cities offers a more suitable habitat than the surrounding rural areas (it is thought that the tree requires a mean annual temperature of 8 degrees Celsius to grow well, which limits its spread to more northern and higher altitude areas). For example, one study in Germany found the tree of heaven growing in 92% of densely populated areas of Berlin, 25% of its suburbs and only 3% of areas outside the city altogether. In other areas of Europe this is not the case as climates are mild enough for the tree to flourish. It has colonized natural areas in Hungary, for example, and is considered a threat to biodiversity at that country's Aggtelek National Park.

Several species of Lepidoptera utilise the leaves of ailanthus as food, including the Indian moon moth (*Actias selene*) and the common grass yellow (*Eurema hecabe*). In North America the tree is the host plant for the ailanthus webworm (*Atteva aurea*), though this ermine moth is native to Central and South America and originally used other members of the mostly tropical Simaroubaceae as its hosts. In its native range *A. altissima* is associated with at least 32 species of arthropods and 13 species of fungi.

Due to the tree of heaven's weedy habit, landowners and other organisations often resort to various methods of control in order to keep its populations in check. For example, the city of Basel in Switzerland has an eradication program for the tree. It can be very difficult to eradicate, however. Means of eradication can be physical, thermal, managerial, biological or chemical. A combination of several of these can be most effective, though they must of course be compatible. All have some positive and negative aspects, but the most effective regimen is generally a mixture of chemical and physical control. It involves the application of foliar or basal herbicides in order to kill existing trees, while either hand pulling or mowing seedlings in order to prevent new growth.

Uses

A male ailanthus silkmoth

In addition to its use as an ornamental plant, the tree of heaven is also used for its wood and as a host plant to feed silkworms of the moth *Samia cynthia*, which produces silk that is stronger and cheaper than mulberry silk, although with inferior gloss and texture. It is also unable to take dye. This type of silk is known under various names: "pongee", "eri silk" and "Shantung silk", the last name being derived from Shandong Province in China where this silk is often produced. Its production is particularly well known in the Yantai region of that province. The moth has also been introduced in the United States.

The pale yellow, close-grained and satiny wood of ailanthus has been used in cabinet work. It is flexible and well suited to the manufacture of kitchen steamers, which are important in Chinese cuisine for cooking mantou, pastries and rice. Zhejiang Province in eastern China is most famous for producing these steamers. It is also considered a good source of firewood across much of its range as it is moderately hard and heavy, yet readily available. The wood is also used to make charcoal for culinary purposes. There are problems with using the wood as lumber, however. Because the trees exhibit rapid growth for the first few years, the trunk has uneven texture between the inner and outer wood, which can cause the wood to twist or crack during drying. Techniques have been developed for drying the wood so as to prevent this cracking, allowing it to be commercially harvested. Although the live tree tends to have very flexible wood, the wood is quite hard once properly dried.

Cultivation

Tree of heaven is a popular ornamental tree in China and valued for its tolerance of difficult growing conditions. It was once very popular in cultivation in both Europe and North America, but this popularity dropped, especially in the United States, due to the disagreeable odor of its blossoms and the weediness of its habit. The problem of odor was previously avoided by only selling pistillate plants since only males produce the smell, but a higher seed production also results. Michael Dirr, a noted American horticulturalist and professor at the University of Georgia, reported meeting, in 1982, a grower who could not find any buyers. He further writes (his emphasis):

> "For most landscaping conditions, it has *no* value as there are too many trees of superior quality; for impossible conditions this tree has a place; selection could be made for good habit, strong wood and better foliage which would make the tree more satisfactory; we once talked with an architect who tried to buy *Ailanthus* for use along polluted highways but could not find an adequate supply".

In Europe, however, the tree is still used in the garden to some degree as its habit is generally not as invasive as it is in America. In the United Kingdom it is especially common in London squares, streets, and parks, though it is also frequently found in gardens of southern England and East Anglia. It becomes rare in the north, occurring only infrequently in southern Scotland. It is also rare in Ireland. In Germany the tree is commonly planted in gardens. The tree has furthermore become unpopular in cultivation in the west because it is short-lived and that the trunk soon becomes hollow, making trees more than two feet in diameter unstable in high winds.

A few cultivars exist, but they are not often sold outside of China and probably not at all in North America:

- 'Hongye' – The name is Chinese and means "red leaves". As the name implies it has attractive vivid red foliage.
- 'Thousand Leaders'.
- 'Metro' – A male cultivar with a tighter crown than usual and a less weedy habit.
- 'Erythrocarpa' – The fruits are a striking red.
- 'Pendulifolia' – Leaves are much longer and hang elegantly.

Traditional Medicine

Nearly every part of *A. altissima* has some application in Chinese traditional medicine. One of the oldest recipes, recorded in a work from AD 732, is used for treating mental illness. It involved chopped root material, young boys' urine and douchi. After sitting for a day, the liquid was strained out and given to an ill person over the course of several days.

Another source from 684 AD, during the Tang dynasty and recorded in Li Shizhen's *Compendium of Materia Medica,* states that when the leaves are taken internally, they make one incoherent and sleepy, while when used externally, they can be effectively used to treat boils, abscesses and itches. This formula calls for young leaves of ailanthus, catalpa and peach tree to be crushed together and the resulting liquid applied to the scalp to stimulate hair growth.

The dried bark, however, is listed in the modern Chinese materia medica as *chun bai pi,* meaning "white bark of spring". Modern works treat it in detail, discussing chemical constituents, how to identify the product and its possible uses. It is prepared by felling the tree in fall or spring, stripping the bark and then scraping off the hardest, outermost portion, which is then sun-dried, soaked in water, partially re-dried in a basket and finally cut into strips. The bark is said to have cooling and astringent properties. It is only prescribed in amounts between 4 and 10 grams, so as not to poison people. Li's Compendium has 18 recipes that call for the bark. A tincture of the root-bark has been claimed to have been used successfully by American herbalists in the 19th century. It contains phytochemicals, such as quassin and saponin, and ailanthone. It is available in most shops dealing in Chinese traditional medicine. The samaras are also used in modern Chinese medicine under the name *feng yan cao,* meaning "herbal phoenix eye".

The plant may be mildly toxic. The noxious odours have been associated with nausea and headaches, as well as with contact dermatitis reported in both humans and sheep, which developed weakness and paralysis. It contains a quinone irritant, 2,6-dimethoxybenzoquinone, as well as quassinoids.

Smooth Brome

Smooth Brome is 2¼–3½' tall and usually unbranched. The culms are light green, hairless or short-pubescent, and terete. The blades of alternate leaves are up to 12" long and 16 mm. (2/3") across; they are greyish blue on the upper side, green on the lower side, hairless, and flat. The leaf blades are ascending to spreading and rather floppy in habit. The leaf sheaths are greyish blue, hairless, and closed, although they usually split open near the ligules. Each culm terminates in a panicle of floral spikelets about 4-8" long; the entire panicle has a tendency to lean sideways or droop from the weight of the spikelets. The branchlets of this panicle are ascending to spreading during the period of bloom, otherwise they are more erect and contracted. The branchlets occur in whorls along the central stalk of the inflorescence, and they often divide into secondary branchlets. These branchlets are green, hairless, and somewhat wiry. Each elongated spikelet is about ¾–1¼" (2-3 cm.) in length; it consists of a pair of glumes and 5-10 fertile lemmas that are arranged in 2 overlapping ranks.

During the period of bloom, the spikelets are whitish green, sometimes with purplish or reddish tints. The glumes are about 5-7 mm. in length; they are elliptic-lanceolate, convex to slightly keeled along their outer surfaces, hairless, and shorter than the lemmas. The individual lemmas are about 8-12 mm. in length; they are elliptic-lanceolate, convex along their outer surfaces, hairless, and usually awnless. On rare occasions, some plants may have lemmas with awns up to 2 mm. in length. The outer surface of each lemma has several longitudinal veins. The blooming period occurs during early to mid-summer. Shortly afterwards, the spikelets and culms become light tan. Disarticulation of the spikelets is above the glumes. The elongated grains are about 8 mm. long and light tan or brown. The root system is fibrous and rhizomatous. Dense clonal colonies of plants are often formed from the rhizomes.

- Cultivation: This robust grass prefers mesic conditions, full or partial sun, and a fertile loam or clay-loam. It can spread aggressively.
- Range & Habitat: Smooth Brome is a common grass that occurs in every county of Illinois. It was introduced as a pasture or forage grass from Eurasia. Habitats include railroad prairies, woodland edges, savannas, pastures, fallow fields, grassy areas along roads, weedy meadows, little-mowed areas of city parks, and waste areas. Smooth Brome has the capacity to invade native prairies and savannas to a greater extent than other Brome grasses (Bromus spp.) from

Eurasia. If anything, its population within the state continues to increase. This grass is found primarily in degraded natural areas and habitats with a history of human-related disturbance, where it can form dense colonies that exclude other plant species.

The majority of brome grasses (Bromus spp.) have been introduced from Eurasia (either deliberately or accidentally). Many of these species are weedy and aggressive, and Smooth Brome (Bromus inermis) is no exception. It can be distinguished from other brome grasses by the absence of awns on its spikelets (rarely with awns up to 2 mm. long); almost all other brome grasses in Illinois have spikelets with conspicuous awns (greater than 2 mm. long). An exception is Rescue Grass (Bromus catharticus), which has been collected only once in Champaign County, Illinois. Rescue Grass is occasionally planted in the southern states as a source of winter forage. It is an annual grass with leaf blades up to ¼" (6 mm.) across. In contrast, Smooth Brome is a perennial grass with leaf blades greater than ¼" (6 mm.) across. Another common name of Bromus inermis is Hungarian Brome.

Ecological Adaptions

Smooth brome was introduced from Europe and has been used in range seedings in many areas of the country. Smooth brome is a good forage producer on the mountain loam sites, and it is also adapted to the upland sites. In the lower rainfall areas, some irrigation is required for optimum production. The elevation range for this grass varies from 3,000 to 12,000 feet. Smooth brome comes on fairly early in the spring, but is not as early as intermediate wheatgrass with which it compares very closely in its adaptations.

Due to it's aggressiveness, the species can be considered weedy. It is not tolerant of prolonged flooding.

- Soils: It is best adapted to fertile, loamy, deep soils including stony loams where there is at least 16 inches of rainfall annually, or supplemental irrigation equivalent to that. Smooth brome is mildly alkaline and moderately salt tolerant.

- Associated Species: Slender wheatgrass, intermediate wheatgrass, aspen, yarrow, big sagebrush.

Uses and Management

Smooth brome is excellent forage for all classes of domestic livestock and for wildlife. Its quality and palatability rapidly decline after inflorescence development. Regrowth in the fall can furnish valuable fall grazing. Smooth brome can withstand heavy use, but produces best with 60% use. If it is grazed too closely for too long it becomes "sod-bound" and unproductive.

Smooth brome is excellent for the control of soil erosion and is used widely for rehabilitation of rangeland for purposes including forage production, wildlife habitat, landscape beautification, cover for recreational areas and campgrounds, roadside seedings, and for watershed stabilization.

Fertilization with nitrogen fertilizer or interseeding with nitrogen fixing legumes is recommended, especially for older smooth brome stands. Recently, it has been determined to be very competitive with desirable native species, so it is often not recommended for restoration seed when other seed is available.

Slender Rush

Slender rush is a clump-forming, grass- or sedge-like perennial, characterized by its wiry dark green stems and terminal clusters of greenish-brown flowers. This species reproduces by seeds.

Root System

The roots are fibrous. Plants sometime produce short rhizomes (horizontal underground stems).

Stems

Slender rush has erect, wiry, rounded, hollow, dark green stems that grow 4 to 24 inches tall.

Leaves

Narrow, grass-like leaves are produced along the lower portion of the stem. The leaf blade (free part of the leaf) is half as long as to longer than the stem, and can be flat or have inwardly rolled edges. The leaf sheaths (part of the leaf surrounding the stem) have whitish, membranous margins, and cover the lower half of the stem. The sheath margins extend upward beyond the top of the sheath to form a pair of whitish, membranous, rounded, flap-like appendages (auricles) (1/25 to 1/6 inch long).

Flowers

Small, greenish-brown flowers are clustered near the tips of branches in a flat-topped, branched flower head. Branches can be 1/3 to 3 inches long. Each flower has 3 scale-like petals and 3 sepals (modified leaves directly below the petals). A pair of narrow, curly, modified leaves (bracts) (1/3 to 4 inches long) usually extends above the flower head.

Fruits and Seeds

Seed pods are egg-shaped, brown and divided into 3 sections, each section containing several seeds. Six scale-like, pointed bracts (modified leaves) surround each seed pod. Seeds are very small (about 1/8 inch long), translucent, orangish-brown, and sometimes winged along one side.

Similar Species

Slender rush may be confused with both grasses and sedges, but it can be distinguished by its round stem, flowers that consist of petals and sepals, and its many-seeded seed pod which is surrounded by 6 pointed bracts.

Biology

Flowering occurs between June and August. Control of slender rush populations may be aided by plowing the infested area, then seeding with a cultivated crop. If the area cannot be cultivated, it could be fertilized, harrowed, and seeded with a mixture containing white clover.

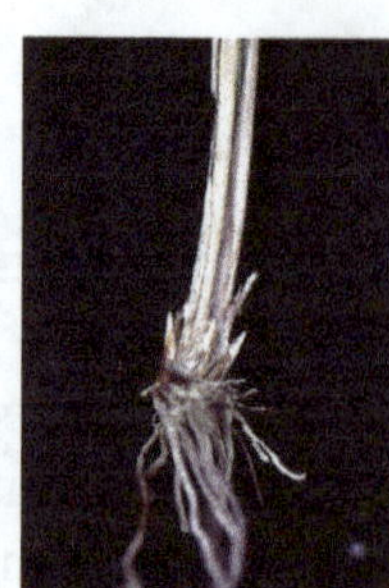

Spotted Knapweed

Spotted knapweed (Centaurea biebersteinnii) is a tap-rooted herbaceous perennial plant that spreads by seed. Spotted knapweed tolerates a wide range of temperature and moisture conditions, and is well adapted to open-forest, pastures. Disturbed sites including transportation corridors, gravel pits, campgrounds, trails, stream and river corridors, and industrial areas are highly susceptible to invasion. Once established on disturbed sites the weed readily moves into adjoining lands. The plant threatens long-term productivity of grazing lands and natural areas by reducing biodiversity and increasing soil erosion. The plant begins growth in spring from basal leaves which form rosettes on a central crown. In early summer the plant will begin to bolt (send up a flower stalk) where the flowers will develop. Flowers occur from midsummer through early September.

Early detection and control is critical to management of spotted knapweed since it spreads easily by water, animals, humans, and vehicles.

Identification

Spotted knapweed is herbaceous broadleaf that grows 8 inches to 4 feet tall and can live at least 9 years. Basal rosette leaves are deeply divided into lobes on both sides of the center vein. Flowering stems are slender, many-branched and have a single flower at the tip of each branch. Flower color is usually pinkishpurple, but can also be light purple or white. Each flower head has bracts located under flower petals that are marked with fine vertical streaks and tipped with a dark comb-like fringe. These bracts give a "spotted" appearance to the flower head.

Reproduction and Spread

Spotted knapweed spreads by seed, and produces about 1000 seeds or more per plant. Seeds remain viable in soil for more than eight years. Spotted knapweed spreads by expansion of existing stands and long-distance movement. Seeds are dispersed by wildlife, livestock, humans, and vehicles, as contaminants in crop seed and hay, and by water flowing in streams, rivers and irrigation canals.

Management

Spotted knapweed can be controlled with various management techniques. Preventing movement of knapweed seeds to non-infested sites is a critical management consideration. People and their motorized vehicles are a major cause of knapweed spread. It is imperative to wash the undercarriage of vehicles that have been in weed-infested areas. Do not walk, drive, or trail livestock through knapweed-infested sites. Only certified weed-free seed, hay and topsoil should be purchased. Livestock should not be grazed in knapweed infested sites during flowering and seed dispersal. Livestock grazing in infested pastures should be held for seven days prior to moving to non-infested pastures. Once knapweed is established, infestations should be controlled as soon as possible to prevent seed production and spread. High risk sites for invasion include transportation corridors, riparian areas, campgrounds, trailheads and other disturbed sites. These areas should be monitored annually for newly invading plants.

Physical/Mechanical/Cultural

Plant communities that maintain competitive desirable vegetation can slow establishment and spread of spotted knapweed. In areas where desirable species are absent, long-term control of spotted knapweed is unlikely. Establishing competitive desirable plants is essential for successfully managing spotted knapweed.

Spotted knapweed is a tap-rooted plant that can be controlled with cultivation to a depth of 7 inches, or hand removal. Persistent pulling or digging can control spotted knapweed if the upper 3 inches of the crown portion of the plant are removed before the plant produces seed. Plants that are pulled after flowering begins should be carefully disposed of to destroy viable seed. While pulling or digging is effective on single plants or relatively small infestations, it is not economically or physically feasible on large, well-established knapweed infestations. Disturbance caused by cultivation or hand-removal increases susceptibility of the site to reinvasion by knapweed or other weeds. Disturbed sites should be seeded to desirable vegetation to stop reinvasion.

Mowing will not control spotted knapweed; however, it can be used to reduce seed production. The optimum time to mow spotted knapweed to reduce seed production is at the late bud growth stage. Mowing in combination with an herbicide treatment may improve control over the herbicide treatment alone.

A single, low-intensity fire does not control knapweed, and may increase knapweed cover and density. Under wildfire situations, spotted knapweed has been observed to increase probably as a result of increased open areas which promote establishment and spread of the knapweeds. Fire followed by herbicide treatments may increase effectiveness of herbicide treatments on knapweed.

Low to moderate use of spotted knapweed by cattle, sheep, goats and some wildlife has been reported. Controlled, repeated grazing of spotted knapweed by sheep can reduce the number of one- and two-year-old spotted knapweed plants within an infestation. Grazing should be timed so that associated grasses are dormant to limit impact on desirable species.

Biological Control

Although most of the insects attack both knapweeds, those most effective on spotted knapweed are described below:

- Two flower head flies (Urophora affinis and U. quadrifasciata). The larvae of these flies induce galls in spotted flower heads which reduce seed production. Seed reductions in excess of 50 in areas where the two fly species coexist.
- A flower head moth (percent are occurring Metzneria paucipunctella). The larvae feed on the florets and seeds of spotted knapweed. Each larva destroys about eight seeds per flower head.

- A root moth (Agapeta zoegana) and a root weevil (Cyphocleonus achates), Larvae of the moth girdle knapweed roots, while the weevil larvae feed in the center of the roots.
- A root beetle (Sphenoptera jugoslavica). The larvae feed in the center of the root. The beetle primarily attacks the roots of diffuse knapweed, but will also attack spotted knapweed.
- Two flower head weevils (Larinus minutus and L. obtusus). The larvae of both weevils feed on knapweed seeds, and the adults feed on knapweed leaves.

Herbicides

Several herbicides that are selective for broadleaf plants can effectively control spotted knapweed. Herbicides containing clopyralid, or aminopyralid are most effective. Dicamba and 2,4-D will provide inconsistent control and must be applied annually for at least 2 years to control mature and seedling plants. Although glyphosate (many trade names) will kill spotted knapweed, it is a non-selective herbicide that will kill both grasses and broadleaf plants. Time of herbicide application depends on the herbicide selected and growing conditions. In general, spotted knapweed can be controlled from spring to early summer before the weed flowers, and during fall if there is adequate moisture.

Weed Management

Since weeds are so prevalent in many areas of the landscape, management techniques are necessary to maintain order. Weed management is most successful when it involves an integrated approach using a variety of methods. The common methods used to manage weeds include prevention and cultural, mechanical, biological, and chemical means.

Prevention

Preventative methods are used to stop the spread of weeds. Preventing the introduction of weeds is usually easier than controlling them after establishment. Preventative practices include cleaning tillage and harvesting equipment of weed seeds and vegetative structures; planting certified, weed-free crop seed; and controlling weeds in barnyards, around structures, and along fencerows, roadways, and ditch banks.

Cultural

Cultural and crop management techniques provide a healthy crop to best compete with weeds. Crop competition can be an inexpensive and effective aid to weed management if used to its fullest advantage. Examples of cultural techniques include following soil test recommendations for fertilizer and lime; selecting the best crop varieties; planting dense crop populations at the proper timing; scouting fields regularly for weeds, insects, and diseases and controlling them when necessary; and including crop rotations in the system. Composting, ensiling, or feeding weeds or weed-infested crops to livestock can destroy the viability of weed seeds. The heat and digestive acids break down the majority of weed seeds. However, some seeds pass through livestock unharmed and can germinate if spread back onto the land.

Preventing weed spread includes controlling weeds around barns and along fences, roads, ditches and woodlands

Mechanical

Mechanical or physical techniques either destroy weeds or make the environment less favorable for seed germination and weed survival. These techniques include hand-pulling, hoeing, mowing, plowing, disking, cultivating, and digging. Mulching (straw, wood chips, gravel, plastic, etc.) can also be considered a mechanical control means since it uses a physical barrier to block light and impede weed growth.

Tine weeders and cultivators can be used to control weed seedlings

Biological

Biological weed control involves the use of other living organisms, such as insects, diseases, or livestock, for the management of certain weeds. In theory, biological control is well suited for an integrated weed management program. However, the limitations of biological control are that it is a long-term under- taking, its effects are neither immediate nor always adequate, only certain weeds are potential candidates, and the rate of failure for past biological control efforts has been fairly high. There have been a few success stories of weed species being managed with insect or disease biocontrol agents. Herbivores such as sheep and goats can provide successful control of some common pasture weeds. Research continues in this area of weed management.

Chemical

Herbicides can be defined as crop- protecting chemicals used to kill weedy plants or interrupt normal plant growth. Herbicides provide a convenient, economical, and effective way to help manage weeds. They allow fields to be planted with less tillage, allow earlier planting dates, and provide additional time to perform the other tasks that farm or personal life require.Without herbicide use, no-till agriculture becomes impossible. However, herbicide use also carries risks that include environmental, ecological, and human health effects. It is important to understand both the benefits and disadvantages associated with chemical weed control before selecting the appropriate control.

Herbicides may not be a necessity on some farms or landscape settings, but without the use of chemical weed control, mechanical and cultural control methods become that much more important. There are many kinds of herbicides from which to choose. Many factors determine when, where, and how a particular herbicide can be used most effectively. Understanding some of these factors enables you to use herbicides to their maximum advantage.

Herbicides

Herbicides can be classified several ways, including by weed control spectrum, labeled crop usage, chemical families, mode of action, application timing/method, and others. Herbicides will be grouped according to mode and site of action, which are also important in understanding herbicide resistance in weeds.

Contact herbicides kill only the plant parts contacted by the chemical, where-as systemic herbicides are absorbed by the roots or foliage and translocated (moved) throughout the plant. Herbicide activity can be either selective or non-selective. Selective herbicides are used to kill weeds without significant damage to desirable plants. Nonselective herbicides kill or injure all plants present if applied at an adequate rate.

Herbicides provide a convenient, economical, and effective way to control weeds

Herbicide Mode and Site of Action

To be effective, herbicides must:

- Adequately contact plants.
- Be absorbed by plants.
- Move within the plants to the site of action without being deactivated.
- Reach toxic levels at the site of action.

The term "mode of action" refers to the sequence of events from absorption into plants to plant death, or, in other words, how an herbicide works to injure or kill the plant. The specific site the herbicide affects is referred to as the "site or mechanism of action." Understanding herbicide mode of action is helpful in knowing what groups of weeds are killed, specifying application techniques, diagnosing herbicide injury problems, and preventing herbicide-resistant weeds.

A common method of grouping herbicides is by their mode of action. Although a large number of herbicides are available in the marketplace, several have similar chemical properties and herbicidal activity. Herbicides with a common chemistry are grouped into "families." Also, two or more families may have the same mode of action, and thus can be grouped into "classes."

Plant Growth Regulators

These herbicides are effective on annual and perennial broadleaf plants and usually have no activity on grasses or sedges, except at high application rates. They produce responses similar to those of natural, growth-regulating substances called auxins. Application of artificial auxins, such as 2,4-D, upsets normal growth as follows:

- Cells of leaf veins rapidly divide and elongate, while cells between veins cease to divide. This results in long, narrow, strap-like young leaves.
- Water content increases, making treated plants brittle and easily broken.
- Cell division and respiration rates increase, and photosynthesis decreases. Food supply of treated plants is nearly exhausted at their death.
- Roots of treated plants lose their ability to take up soil nutrients, and stem tissues fail to move food effectively through the plant.

The killing action of growth-regulating chemicals is not caused by any single factor but results from the effects of multiple disturbances in the treated plant.

Injury Symptoms

Broadleaf plant leaves become crinkled, puckered, strap shaped, stunted, and malformed; leaf veins appear parallel rather than netted, and stems become crooked, twisted, and brittle, with shortened internodes. If injury occurs in grasses (e.g., corn), new leaves do not unfurl but remain tightly rolled in onion-like fashion, and stems become brittle, curved, or crooked, with short internodes. A lesser effect in corn is the fusion of brace roots, noticed later in the season.

Amino Acid Biosynthesis Inhibitors

These herbicides are effective mostly on annual broadleaves, while a few in this large group have activity on grasses, nutsedge, and/or perennial plants. (Glyphosate [Roundup], for example, is a broad-spectrum herbicide and has activity on all types of plants.) These herbicides work by interfering with one or more key enzymes that catalyze the production of specific amino acids in the plant. When a key amino acid is not produced, the plant's metabolic processes begin to shut down. The effect is like that of an assembly line worker not doing his or her job. Different herbicides affect different enzymes that catalyze the production of various amino acids, but the result is generally the same--the shutdown of metabolic activity with eventual death of the plant.

Injury Symptoms

Plants that are sensitive to these herbicides stop growth almost immediately after foliar treatment; seedlings die in two to four days, established perennials in two to four weeks. Plants become straw colored several days or weeks after treatment, gradually turn brown, and die.

Fatty Acid (Lipid) Biosynthesis Inhibitors

These herbicides are rapidly absorbed by grasses and are translocated to the growing points, where they inhibit meristematic activity, stopping growth almost immediately. They have no activity on broadleaf plants and are most effective on warm-season grasses such as Johnsongrass, shattercane, corn, fall panicum, giant foxtail, and crabgrass. Cool-season grasses such as quackgrass, annual and perennial ryegrass, orchardgrass, timothy, and small grains are not as sensitive as the warm-season grasses. Some of these herbicides are weaker on perennial species than other products. They are frequently referred to as "post-grass" herbicides.

Injury Symptoms

Growing points are killed first, resulting in the death of the leaves' inner whorl. Older, outer leaves of seedlings appear healthy for a few days, and those of perennials for a couple of weeks, but eventually they also wither and die. After several weeks, the growing points begin to rot, allowing the inner leaves to be pulled out of the whorl. Sensitive grasses commonly turn a purplish color before dying.

Seedling Growth Inhibitors (Root and Shoot)

Herbicides in this group prevent cell division primarily in developing root tips and are effective only on germinating, small-seeded annual grasses and some broadleaves.

Injury Symptoms

Seeds of treated broadleaved plants germinate, but they either fail to emerge or emerge as severely stunted seedlings that have thickened, shortened lower stems, small leaves, and short, club- shaped roots. Seedlings of tap-rooted plants, such as soybeans and alfalfa, are usually not affected, nor are established plants with roots more than a couple of inches deep.

Grass seeds germinate, but generally fail to emerge. Injured seedlings have short, club-shaped roots and thickened, brittle stem tissue. Seedlings die from lack of moisture and nutrients because of the restricted root system.

Seedling Growth Inhibitors (Shoot)

Herbicides in this class are most ef- fective on annual grasses and yellow nutsedge. They are sometimes referred to as "pre-grass" herbicides. Depending on the product, some will control small-seeded annual broadleaves. These herbicides cause abnormal cell development or prevent cell division in germinating seedlings. They stop the plant from growing by inhibiting cell division in the shoot and root tips while permitting other cell duplication processes to continue. Then follows a slow decline in plant vigor.

Injury Symptoms

Germinating grasses normally do not emerge. If they do, young leaves fail to unfold, resulting in leaf looping and an onion-like appearance. The tip of the terminal leaf becomes rigid, not free flapping (flag like). The leaves of broad-leaved plants turn dark green, become wrinkled, and fail to unfold from the bud. The roots become shortened, thickened, brittle, and club like.

Photosynthesis Inhibitors (Mobile)

These herbicides are effective primarily on annual broadleaves, while some provide control of grasses as well. Photosynthesis-inhibiting herbicides block the photosynthetic process so captured light cannot be used to produce sugars. In the presence of light, green plants produce sugar from carbon dioxide and water. Energy is needed for carbon, hydrogen, and oxygen atoms to rearrange and form sugar. To supply this necessary energy, electrons are borrowed from chlorophyll (the green material in leaves) and replaced by electrons split from water. If chlorophyll electrons are not replaced, the chlorophyll is destroyed and the plant's food manufacturing system breaks down. The plant slowly starves to death due to lack of energy.

As soil-applied treatments, these herbicides permit normal seed germination and seedling emergence, but cause seedlings to lose their green color soon afterward. With the seeds' food supply gone, the seedlings die. These herbicides are more effective on seedling weeds than on established perennial weeds. Herbicides such as prometon (Primitol) and tebuthiuron (Spike) are considered soil sterilants. Soil sterilants are nonselective chemicals that can kill existing vegetation and keep the soil free from vegetation for one or more years.

Injury Symptoms

In broadleaved plants, early seedling growth appears normal, but shortly after emergence (when energy reserves in cotyledons are depleted), leaves become mottled, turn yellow to brown, and die. In most cases, the oldest leaves turn yellow on the leaf margins first, the veins remain green, and eventually the plant turns brown and dies. Herbaceous and woody perennials starve very slowly because they have large energy reserves in roots or rhizomes to live on while photosynthesis is inhibited. The herbicide may have to effectively inhibit photosynthesis for a full growing season to kill trees or brush. This kind of death may be slow, but it is certain.

Photosynthesis Inhibitors (Nonmobile; Rapid-Acting)

Herbicides in this group have activity on primarily annual and some perennial broadleaves and are applied to the plant foliage. The mode of action is the same as the mobile photosynthesis inhibitors.

Injury Symptoms

Their activity within the plant is similar to that of the mobile photosynthesis inhibitors, except the injury occurs at the site of contact, causing "leaf burning" and eventual death of the plant.

Cell Membrane Disrupters

These herbicides control mostly broadleaves. Certain products have some activity on grasses, and paraquat (Gramoxone) provides broad-spectrum control of many different species. These herbicides are referred to as contact herbicides and they kill weeds by destroying cell membranes. They appear to burn plant tissues within hours or days of application. Good coverage of the plant tissue and bright sunlight are necessary for maximum activity. The activity of these herbicides is delayed in the absence of light.

Injury Symptoms

All contact herbicides cause cellular breakdown by destroying cell membranes, allowing cell sap to leak out. Effected plants initially have a "water-soaked" appearance, followed by rapid wilting and "burning," or leaf speckling and browning. Plant death occurs within a few days.

Pigment Inhibitors

These herbicides provide control of many annual broadleaves and some grasses. These products are referred to as "bleachers" since they inhibit carotenoid biosynthesis or the HPPD enzyme by interfering with normal chlorophyll formation.

Injury Symptoms

Symptoms are very evident and easy to identify. Effected plants either do not emerge or emerge white or bleached and eventually die. Older leaf tissue is affected first.

Phosphorylated Amino Acid (Nitrogen Metabolism) Disrupters

This herbicide provides broad-spectrum control of most annual grasses and broadleaves and some perennials. It affects growth by disrupting nitrogen metabolism, thus interfering with other plant processes. It is a contact herbicide with slight translocation throughout the plant. Good spray coverage and sunlight are important for maximum efficacy.

Injury Symptoms

Injury is similar to that of the cell membrane disrupter herbicides. Sensitive plants show "leaf burning," yellowing and browning, and eventual death after a week or so. Perennials generally take longer for symptoms and death to occur.

Unknown Herbicides

This category contains miscellaneous products for which the mode of action and family are unknown. Dazomet (Basamid) and metam (Vapam) are considered soil fumigants. These products are applied to the soil and covered with a gas-tight tarp; there, they are converted to gases and penetrate the soil to kill weeds, diseases, and nematodes. Endothall (Aquathol) is used for aquatic weed control. Fosamine (Krenite) is used in noncrop areas to control perennial weeds and brush. Other compounds such as pelargonic acid (Scythe), a fatty-acid herbicide, and clove oil and vinegar are contact, nonselective, broad-spectrum, foliar-applied products that are sometimes used for weed control in organic crop production settings. However, because they basically "burn" only the plant tissue they contact, there is potential for plant regrowth.

Selective herbicides control weeds without causing injury to the crop or other desirable plants

Herbicide Resistance

A number of weed species that were once susceptible to and easily managed by certain herbicides have developed resistance. These weeds are no longer controlled by applications of previously effective herbicides. As a result of repeatedly using a certain type of herbicide on the same land, many different species of weeds have developed resistance to these chemicals. Currently, about 180 weed species (more than 300 weed biotypes) worldwide are resistant to about ten different herbicide families. It is believed that within any population of weeds, a few plants have sufficient tolerance to survive any herbicide that is used. Since only the survivors can produce seed, it is only a matter of time until the population of resistant weeds outnumbers the susceptible type. Depending on the herbicide family and weed species, resistance can occur within 5 to 20 years. Certain precautions, such as tank-mixing, crop rotations, and a combination of weed management techniques, must be taken to prevent resistance.

Growers, consultants, and those working with herbicides to manage weeds should know which herbicides are best suited to combat specific resistant weeds. The Weed Science Society of America (WSSA) developed a grouping system to help with this process. Herbicides that are classified as the same WSSA group number kill weeds using the same mode of action. WSSA group numbers can be found on many herbicide product labels and can be used as a tool to choose herbicides in different mode of action groups so mixtures or rotations of active ingredients can be planned to better manage weeds and reduce the potential for resistant species.

Times of Application

The following terms describe herbicides based on when they are applied:

- Preplant incorporated: applied to soil and mechanically incorporated into the top 2 to 3 inches of soil before the crop is planted.
- Preplant: applied to soil before the crop is planted.

- Preemergence: applied after the crop is planted but before it emerges.
- Postemergence: applied after crop emergence.

Although these terms normally refer to application in relation to crops, they may also imply application in relation to weeds. Always be certain whether reference is being made to the crop or to the weed. In no-till situations, it is possible for an herbicide application to be preplant or preemergence to the crop but postemergence to weeds. Some herbicides must be preplant or preemergence to the weed for maximum activity.

Methods of Application

The following terms refer to the ways herbicides can be applied:

- Broadcast: applied over the entire field.
- Band: applied to a narrow strip over the crop row.
- Directed: applied between the rows of crop plants with little or no herbicide applied to the crop foliage.
- Spot treatment: applied to small, weed-infested areas within a field.

Drift

Drift is the movement of any pesticide through the air to areas not intended for treatment. During application, droplet or particle drift occurs as spray droplets or dust particles are carried by air movement from the application area to other places. Vapor drift takes place after application as herbicides evaporate (volatilize) and yield fumes (gases) are carried on wind currents and deposited on soils or plants in untreated areas.

Drift may injure sensitive crops, ornamentals, gardens, livestock, wildlife, or people and may contaminate streams, lakes, or buildings. It may contaminate crops and cause illegal or intolerable residues. Excessive drift may mean poor performance in the desired spray area because the application rate is lower than expected.

Highly active chemicals present the greatest drift hazard because extremely small amounts can cause severe problems. For example, growth regulator herbicides such as 2,4-D, dicamba, and picloram at a rate of 1 ounce per acre can deform sensitive crops such as tobacco, grapes, or tomatoes.

Vapor drift from Command (clomazone) that has not been incorporated can cause bleaching of chlorophyll in sensitive plants within a quarter mile of application. Vapor drift problems can often be avoided by using nonvolatile formulations. Essentially, no vapor drift hazard is involved in the use of amine formulations of 2,4-D. Soil incorporation of Command and a microencapsulated formulation greatly reduces vapor loss of this herbicide.

Particle drift depends on the size of the particle or droplet, and droplet size depends on pressure and nozzle design. Very small particles of fog or mist present the greatest drift hazard. To minimize particle drift, calibrate equipment to create droplets about the size of light rain. Most nozzles can be adjusted to a pressure that permits droplet formation as a result of surface tension. If nozzles are operated at this pressure, a minimum of mist-sized droplets will be formed. For some nozzles, this pressure may be as little as 15 psi; for others, it may be 30 psi.

The distance particles will drift increases with the height of release. Wind velocities usually are lower close to the ground. Therefore, sprays should be released as close to the soil surface or vegetation as adequate coverage permits.

Pesticide drift is influenced by wind, air temperature, boom height, and spray droplet size

Drift hazard usually is minimized if prevailing winds are blowing away from sensitive crops, but a sudden shift in wind direction could result in serious damage. If possible, do not apply pesticides when wind speed is greater than 5 mph.

High temperatures increase the loss of volatile herbicides. Esters of 2,4-D rapidly evaporate at temperatures above 800 °F. The use of such ester formulations should be restricted to fall, winter, and early spring because sensitive plants are not present and lower temperatures reduce vapor drift hazard.

Drift control should be considered with each pesticide application. You can prevent severe drift problems by:

- Using sprayer nozzles especially designed for drift reduction;
- Using low volatile or nonvolatile formulations;
- Using low spray-delivery pressures (15-30 psi) and nozzles with a larger orifice;
- Using drift-inhibiting adjuvants in the spray mixture when spraying under less-than-ideal conditions;

- Using nozzles that allow for lowered boom height;
- Avoiding application of volatile chemicals at high temperatures;
- Spraying when wind speed is low (less than 5 mph) or when the wind is blowing away from areas that should not be contaminated;
- Spraying during the early morning or evening hours when there is usually less wind;
- Leaving border areas unsprayed if they are near sensitive crops.

Persistance

The residual life or length of time an herbicide persists in the soil is the length of time it can be expected to control weeds. Residual toxicity, if not considered, may injure the next crop planted in a herbicide-treated field. Inactivation, breakdown, and disappearance of herbicides are influenced by the following factors.

Microbial Degradation

Microorganisms feed on all types of organic matter, including organic herbicides. Microbial degradation is the primary means of herbicide breakdown. Some herbicides are more readily attacked by microorganisms than others, often because of minor differences in chemical structure that permit rapid decomposition in some cases and block decomposition in others. Soil temperature, aeration, pH levels, organic matter, and moisture levels favorable for microbial growth promote rapid herbicide breakdown. Microbial degradation takes place primarily in the top foot of soil, where microbial activity is the greatest.

Chemical Degradation

Herbicides may be inactivated upon reaction with salts, acids, and other substances in the soil. These reactions are affected by the same environmental factors that influence microbial breakdown. Chemical degradation can occur anywhere in the soil profile and is the primary process responsible for herbicide dissipation below the top foot of soil, where microbial activity is limited or nonexistent.

Runoff

Water moving over the surface of a field or treated area can carry herbicide with it. The greatest loss of herbicide occurs when the herbicide is applied to the soil surface and is washed off by the first rain after application. If the herbicide is incorporated or leached into the soil with light rains or irrigation, most loss occurs only with erosion after the herbicide is adsorbed to soil particles.

Leaching

Water carries herbicides into and ultimately out of the root zone. The portion lost to leaching depends on soil texture, herbicide solubility, and amount and intensity of rainfall. As a rule, herbicides leach most from sandy soils and least from clay soils or soils high in organic matter.

Adsorption

After application, herbicides may become adsorbed (bound) to clay and organic matter particles. The extent of adsorption increases as the percentage of organic matter and clay increases. Adsorption reduces the amount of chemical available to plants and slows leaching. Herbicides are then degraded by various means.

Volatilization

Some herbicides may be rapidly lost as vapors after application. Loss as vapor reduces the persistence of dinitroaniline and thiocarbamate herbicides and Command. The rate of vapor loss is influenced by soil moisture, temperature, and adsorption. Evaporation of herbicides increases as sand content, soil moisture, and soil temperature increase. Incorporation into soil immediately after application reduces this kind of loss.

Photodecomposition

Sunlight may inactivate herbicides- A factor that may contribute to a decline in effectiveness of unincorporated herbicides such as trifluralin (Treflan) and benefin (Balan). Exposure to light for two or more hours reduces the effectiveness of trifluralin and related herbicides and can be avoided by soil incorporation.

Plant Uptake

Herbicides may be absorbed by plant roots or leaves and inactivated within the plant. This effect generally accounts for a relatively small amount of herbicide removal.

Crop Removal

If a crop is harvested or removed from the treated area before rain has washed the herbicide off the foliage or before the plant has had time to metabolize the residue, the herbicide will be removed with the crop. This seldom happens because herbicides are not commonly used close to harvest. However, if grass clippings are collected shortly after treatment and used to mulch a garden, there may be enough herbicide on the grass to damage the garden plants.

References

- Weed, plant: britannica.com, Retrieved 2 April, 2019
- Introduction-to-weeds-and-herbicides: psu.edu, Retrieved 12 February, 2019
- Gr-foxtail, plants, grasses: illinoiswildflowers.info, Retrieved 16 August, 2019
- Ohawkweed, weedlist: idahoweedawareness.net, Retrieved 19 May, 2019
- Aurantiacum, weeds, biodiversity: environment.gov.au, Retrieved 29 April, 2019
- Dodder-weed-control, weeds, plant-problems: gardeningknowhow.com, Retrieved 25 March, 2019
- van Mourik, Thomas A (2007). Striga hermonthica seed bank dynamics process quantification and modelling (PhD thesis). Wageningen University. ISBN 978-90-8504-692-9
- Weeds: dpi.nsw.gov.au, Retrieved 5 June, 2019
- Single-weed, weedguide: oardc.ohio-state.edu, Retrieved 16 June, 2019
- Bull-thistle, plants, weeds: illinoiswildflowers.info, Retrieved 11 July, 2019
- Bull-thistle, weed-identification, noxious-weeds, animals-and-plants, environment: kingcounty.gov, Retrieved 15 February, 2019
- Smooth-brome, plants, grasses: illinoiswildflowers.info, Retrieved 25 March, 2019
- Smooth-brome, grasses-and-grasslikes, rangeplants: extension.usu.edu, Retrieved 27 January, 2019
- Young, Nelson D.; Steiner, Kim E.; dePamphilis, Claude W. (Autumn 1999). "The Evolution of Parasitism in Scrophulariaceae/Orobanchaceae: Plastid Gene Sequences Refute an Evolutionary Transition Series" (PDF). Annals of the Missouri Botanical Garden. 86 (4): 876–93. doi:10.2307/2666173. JSTOR 2666173
- Spotted-knapweed, weeds: wisc.edu, Retrieved 30 August, 2019
- Introduction-to-weeds-and-herbicides: psu.edu, Retrieved 9 January, 2019
- Munz, Philip Alexander; David D. Keck (1973) [1959–1968]. A California Flora and Supplement. Berkeley: University of California Press. p. 993. ISBN 978-0-520-02405-2

Chapter 4

Irrigation Management

Irrigation is a method that involves the application of controlled amount of water to plants at regular intervals. The main methods of irrigation include drip irrigation, surface irrigation, sprinkler irrigation and subsurface drip irrigation. This chapter discusses in detail the methods and types of irrigation as well as their management.

Irrigation

Irrigation is the process of applying water to the crops artificially to fulfil their water requirements. It fulfils the water requirement of crops. Nutrients may also be applied to the crops through irrigation.

The various sources of water for irrigation are wells, ponds, lakes, canals, tube-wells, and even dams. Irrigation offers moisture required for growth and development, germination, and other related functions.

Water moistens the soil and thus helps in penetration of roots even into the dry field. The frequency, rate, amount and time of irrigation are different for different crops and also vary according to soil and seasons. For example, summer crops require a higher rate of water as compared to winter crops.

The Irrigation Canal

Types of Irrigation

There are different types of irrigation practised to improve crop yield. These include:

- Surface Irrigation: In this system, no irrigation pump is involved. Here, water is distributed across the land by gravity.

- Localized Irrigation: In this system, water is applied to each plant through a network of pipes under low pressure.
- Sprinkler Irrigation: Water is distributed from a central location by overhead high-pressure sprinklers or from sprinklers from the moving platform.
- Drip Irrigation: In this type, drops of water are delivered near the roots of the plants.
- Centre Pivot Irrigation: In this, the water is distributed by a sprinkler system moving in a circular pattern.
- Sub-irrigation: Water is distributed through a system of pumping stations gates, ditches and canals by raising the water table.
- Manual Irrigation: This a labour intensive and time-consuming system of irrigation. Here, the water is distributed through watering cans by manual labour.

Modern Irrigation System Planning and Construction

Water Supply

The first consideration in planning an irrigation project is developing a water supply. Water supplies may be classified as surface or subsurface. Though both surface and subsurface water come from precipitation such as rain or snow, it is far more difficult to determine the origin of subsurface water.

In planning a surface water supply, extensive studies must be made of the flow in the stream or river that will be used. If the streamflow has been measured regularly over a long period, including times of drought and flood, the studies are greatly simplified. From streamflow data, determinations can be made of the minimum, maximum, average daily, and average monthly flows; the size of dams, spillways, and downstream channel; and the seasonal and carry-over storage needed. If adequate streamflow data are not available, the streamflow may be estimated from rain and snow data, or from flow data from nearby streams that have similar climatic and physiographic conditions.

The quality, as well as the quantity, of surface water is a factor. The two most important considerations are the amount of silt carried and the kind and amount of salts dissolved in the water. If the silt content is high, sediment will be deposited in the reservoir, increasing maintenance costs and decreasing useful life periods. If the salt concentration is high, it may damage crops or accumulate in the soil and eventually render it unproductive.

Subsurface sources of water must be as carefully investigated as surface sources. In general, less is known about subsurface supplies of water than about surface supplies, so, therefore, subsurface supplies are harder to investigate. Engineers planning a project need to know the extent of the basic geological source of water (the aquifer), as well as the amount the water level is lowered by pumping and the rate of recharge of the aquifer. Often the only way for the engineer to obtain these data reliably is to drill test wells and make on-site measurements. Ideally, a project is planned so as not to use more subsurface water than is recharged. Otherwise, the water is said to be "mined," meaning that it is being used up as a natural resource and its use is considered unsustainable.

Two sources of water not often thought of by the layperson are dew and sewage or wastewater. In certain parts of the world, Israel and part of Australia, for example, where atmospheric conditions are right, sufficient dew may be trapped at night to provide water for irrigation. Elsewhere the supply of wastewater from some industries and municipalities is sufficient to irrigate relatively small acreages. Recently, due to greater emphasis on purer water in streams, there has been increased interest in this latter practice.

In some countries (Egypt for example) sewage is a valuable source of water. In others, such as the United States, irrigation is looked upon as a means of disposing of sewer water as a final step in the wastewater treatment process. Unless the water contains unusual chemical salts, such as sodium, it is generally of satisfactory quality for agricultural irrigation. Where the practice is used primarily as a means of disposal, large areas are involved and the choice of crop is critical. Usually only grass or trees can withstand the year-round applications.

Before a water supply can be assured, the right to it must be determined. Countries and states have widely varying laws and customs that determine ownership of water. If the development of a water supply is for a single purpose, then the determination of ownership may be relatively simple; but if the development is multipurpose, as most modern developments are, ownership may be difficult to determine, and agreements must be worked out among countries, states, municipalities, and private owners.

The area that can be irrigated by a water supply depends on the weather, the type of crop grown, and the soil. Numerous methods have been developed to evaluate these factors and predict average annual volume of rainfall needed. Some representative annual amounts of rainfall needed for cropland in the western United States are 305 to 760 mm (12 to 30 inches) for cereal grains and 610 to 1,525 mm (24 to 60 inches) for forage. In the Near East, cotton needs about 915 mm (36 inches), whereas rice may require two to three times that amount. In humid regions of the United States, where irrigation supplements rainfall, grain crops may require 150 to 230 mm (6 to 9 inches) of water. In addition to satisfying the needs of the crop, allowances must be made for water lost directly to evaporation and during transport to the fields.

Transport Systems

The type of transport system used for an irrigation project is often determined by the source of the water supply. If a surface water supply is used, a large canal or pipeline system is usually required to carry the water to the farms because the reservoir is likely to be distant from the point of use. If subsurface water drawn from wells is used, a much smaller transport system is needed, though canals or pipelines may be used. The transport system will depend as far as possible on gravity flow, supplemented if necessary by pumping. From the mains, water flows into branches, or laterals, and finally to distributors that serve groups of farms. Many auxiliary structures are required, including weirs (flow-diversion dams), sluices, and other types of dams. Canals are normally lined with concrete to prevent seepage losses, control weed growth, eliminate erosion hazards, and reduce maintenance. The most common type of concrete canal construction is by slip forming. In this type of construction, the canal is excavated to the exact cross section desired and the concrete placed on the earth sides and bottom.

Pipelines may be constructed of many types of material. The larger lines are usually concrete whereas laterals may be concrete, cement–asbestos, rigid plastic, aluminum, or steel. Although pipelines are more costly than open conduits, they do not require land after construction, suffer little evaporation loss, and are not troubled by algae growth.

Water Application

After water reaches the farm it may be applied by surface, subsurface, or sprinkler-irrigation methods. Surface irrigation is normally used only where the land has been graded so that uniform slopes exist. Land grading is not necessary for other methods. Each method includes several variations, only the more common of which are considered.

Surface irrigation systems are usually classed as either flood or furrow systems. In the flood system, water is applied at the edge of a field and allowed to move over the entire surface to the opposite side of the field. Grain and forage crops are quite often irrigated by flood techniques. The furrow system is used for row crops such as corn (maize), cotton, sugar beets, and potatoes. Furrows are plowed between crop rows and the water is run in the furrows. In either type of surface irrigation systems, waste-water ditches at the lower edge of the fields permit excess water to be removed for use elsewhere and to prevent waterlogging.

Irrigation sprinklers

Subirrigation is a less common method. An impermeable layer must be located below, but near, the root zone of the crop so that water is trapped in the root zone. If this condition exists, water is applied to the soil through tile drains or ditches.

Sprinklers have been used increasingly to irrigate agricultural land. Little or no preparation is needed, application rates can be controlled, and the system may be used for frost protection and the application of chemicals, such as pesticides, herbicides, and fertilizers. Sprinklers range from those that apply water in the form of a mist to those that apply an inch or more per hour.

Evaporation and Seepage Control

Various techniques have been tried to reduce losses of irrigation water. Two major sources of loss, particularly from surface supplies and surface systems, are evaporation and page from reservoirs and canals. Many studies have been made of techniques to suppress evaporation. One of the more promising appears to be application of a special alcohol film on the surface, which retards evaporation by about 30 percent and does not reduce the quality of the water. The primary problem in its use is that it is fragile; a strong wind can blow it apart and expose the water to evaporation.

Seepage has largely been controlled by lining main and distribution channels with impervious material, typically concrete. Other materials used are asphalt and plastic film, though plastic tends to deteriorate if it is exposed to sunlight.

Typical Systems

The typical surface irrigation system utilizes a publicly developed water supply—e.g., a river-basin reservoir. The public project also constructs the main canals to take water from the reservoir to the agricultural land. In general the canals flow by gravity, but lift stations are often required. Supply and field canals are used to bring the water to the individual field, where it is applied to the land either by furrow or by flooding method.

Until recently most sprinkler-irrigation systems depended on privately developed water supplies, but many modern sprinkler systems have been able to draw on public water supplies. In either case, a pump is required to pump water from a large well (3,785 litres, or 1,000 gallons, per minute or more) or a supply canal. The water goes into the system main and thence to a sprinkler unit. Many automatic or semiautomatic moving sprinkler systems travel over the field applying water. Two common units are the so-called centre pivot and the travelling sprinkler. The centre-pivot unit is anchored at the centre of the field; a long lateral (arm) with sprinklers mounted on it sweeps the field in a circle. The system has the disadvantage of missing the corners of a square field. A travelling sprinkler is mounted on a trailer and propelled across the field in a lane that has been left unplanted. The unit drags a flexible hose connected to the main supply line. When it reaches the end of the lane, it is automatically shut off and can be moved to the next lane. Despite some shortcomings, all sprinkler systems are effective in applying a controlled amount of water at a high level of efficiency with a minimum of labour.

Methods and Types of Irrigation

Surface Irrigation

In Surface Irrigation, water is applied and distributed over the soil surface by the help of gravity. Throughout the world, this is the most commonly used type of irrigation process. Surface Irrigation is an unchanged process and it is older than its recognition. For thousands of years, people are practicing this irrigation process. There is a misconception that, Surface Irrigation process is sometimes considered as Flood Irrigation because of uncontrolled water distribution. But if applied correctly, Surface Irrigation can be your best irrigation process to water your plants.

Working Principle of Surface Irrigation

The surface Irrigation system is very easy yet natural. You need to go through only four phases and your process is done. At first, water is supplied to the top end of the fields, it naturally follows to the field length. Now it is the Advance Phase. It depends on the length of time of the water supply. Again, water is supplied to the top end of the fields, it gradually runs off to the pond. Next is the Storage Phase. The period of time in this is between the end of the advance phase and the shut-off phase. Depletion phase is a short period of time when the length of the field is submerged. Last is the Recession phase. It is the time period while the waterfront goes toward the downstream end of the field.

Types of Surface Irrigation Methods

Free Flood Method

This is a common methods of irrigation of wetland rice where water may be applied as a continuous flood, rotational flood or an intermittent flood. This is a suitable method for an area where cheap and abundant supply of water is available. Flooding consists of opening a water channel in a plot of field, so that water can flow freely in all directions and cover surface of the land in a continuous sheet.

Advantages

1. No other field layout become necessary except the levelling and preparation of impervious boundary bunds.
2. No land area is utilized for water distribution. As a result, wastage of land becomes minimum.

3. Labour supervision is required for application of water in the field.
4. Labour requirement is minimum.

Disadvantages

1. It is the most inefficient methods of irrigation as only about 20 per cent of water is actually used by plants and the rest is being lost as run off, seepage and evaporation.
2. Levelling of land increases cost of cultivation.
3. This method is unsuitable for crops that are sensitive to water logging.
4. Crop growth is not uniform as the water distribution by this method is very uneven.
5. There is a possibility of soil erosion.

Furrow Methods

This method is suitable for row crops like potato cabbage, cauliflower, brinjal, chilli, sugarcane, cotton, tobacco, tomato etc. water is applied in the furrows which are developed between crop rows in the planting and cultivating processes. There is usually me furrow between each row of crop. In orchards and widely spread vegetable crops, several furrows may be opened between the rows of the crop and the water is led simultaneously in all furrows. Furrows, are laid down along the slope when the slope of the land is about three per cent. The furrows are on graded contour when the slope exceeds three per cent and goes upto 15 per cent. The length of furrow varies with soil type, slope of the land and the quantity of water to l applied and may vary from 10 metres to 100 metres.

Furrow method of irrigation

Advantages

1. There is a great advantage in the use of water which is of great consideration in lift irrigation.
2. This method is suitable for crops which are sensitive to saturated soil condition in the root zone.
3. iWater in the furrows contacts only one fifth to one half of the land surface, thereby reducing puddling, crusting of the soil and evaporation losses.
4. Water is applied uniformly by this method.
5. It is especially suitable to crops like maize that are injured by contact with water as water does not come in contact with the plant stem.
6. The method reduces labour requirements in land preparation and irrigation.

Disadvantages

1. Cost involved in the preparation of furrow is more.
2. This method is not suitable for all areas, especially cannel areas.
3. Uniform distribution of water in all furrows is not possible.

Following are the different types of furrow irrigations:

1. Straight furrow method of irrigation: This method is practiced in flat lands.
2. Contour furrow method of irrigation: This method is practiced in sloping land in which the direction of flow is across a slope rather than down slope. The furrows are laid out with a gentle uniform grade to carry the irrigation streams.
3. Corrugated furrow irrigation: Corrugations are small furrows which are used for close growing crops such as beet, cabbage, cauliflower, turnip, etc. In this system more and smaller furrows are utilized and crop rows are not necessarily related to the irrigation furrows. The size and spacing of corrugation vary with the soil. The corrugation should be more closely spaced, generally 45 to 90cm.

Check and Furrow Method

It is a modified form of furrow irrigation. This method suitable for a land which is quite levelled. The water is lead in first furrow and the flow of water is continued to second furrow through a breach in the ridge at the terminal end of the run. After completion of the run in the second furrow, the terminal end of the flow is continued to the third furrow through a breach. A number of furrows are irrigated with a single run of water that moves in a zigzag way through furrows.

Check and furrow method of irrigation

Advantages

1. Water is applied uniformly and the wastage of water minimum. There is less evaporative loss from furrows.
2. Soil gets sufficient time to absorb water.
3. It is a very suitable method for irrigating row crops in level land.
4. Soil erosion is checked.

Disadvantages

1. Uneven distribution of water in the different furrows.
2. The method is only suitable for row crops.
3. Irrigation cost is higher as it involves much labour.
4. More time is required for irrigation.

Border or Check Basin irrigation

The field is divided into smaller unit areas or beds measuring 10 to 100 sq metre or even more. Bunds or ridges are constructed around the areas forming basin within which irrigation water may be controlled. The basins are levelled in both directions. Small irrigation channels are provided between two adjacent rows of beds. Water is conveyed to the field by the stream of supply channel and lateral field channel. The supply channel is aligned in the upper side of the area and there is usually one lateral for every two rows of check basins Water from the lateral channel is turned in the beds and is cut oil when sufficient water has been admitted to the basin. The basins are filled to the desired depth and the water is retained until it infiltrates the soil.

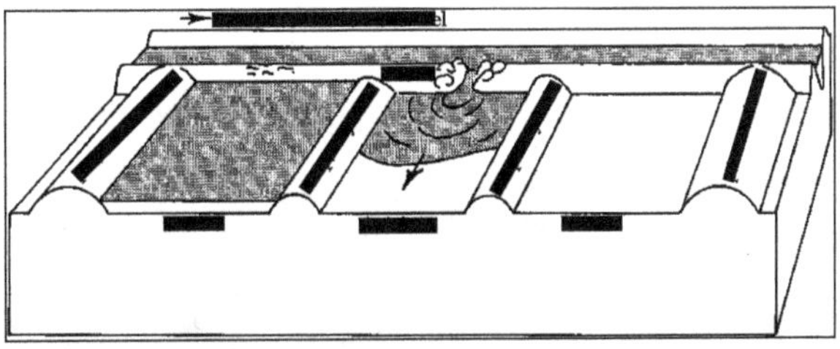

Check basin method of irrigation

This method is especially suitable for irrigating grain and fodder crops in heavy soil where water is absorbed very slowly.

Advantages

1. It is more efficient method of irrigation arid ensures uniform application of water.
2. It has good control on irrigation water and high water application efficiency.
3. Though the Initial cost is high, it requires less labour and has low maintenance cost.
4. Wastage of water and soil erosion is reduced as there is bund around the basin.

Disadvantages

1. There are too many ridges which occupy the larger area of land. It involves the wastage of land.
2. Lay out requirement in land preparation and irrigation is much bigger in check basin irrigation as compared to other methods.
3. Repairing of ridges and careful supervision during irrigation is needed.
4. Uneven distribution of water within the plot.
5. The method impedes surface drainage, precise land grading and shaping are required.

Border Strip Method of Irrigation

The field is divided into a number of long parallel (3-10 ni x 50-30m) strips which is called border. The border is separated with small parallel low ridges on the sides. The individual strips are levelled perfectly and each strip is irrigated independently by turning a stream of water at the upper elevation. The water can flow down the slope with nearly uniform depth. This method is suitable to irrigate all close growing crops like wheat, barley, fodder crops and legumes. But this method is not suitable to irrigate rice which requires standing water during most part of their growing season. This method is suitable to soils having moderately low to moderately high infiltration rate.

Boarder strip method of irrigation

Advantages

1. It is the best method to irrigate close growing crops.
2. Uniform distribution and high water application efficiencies are possible if the system is properly designed.
3. Labour requirement is less to irrigate the field.
4. Operation of this system is simple and easy.
5. Excess rainwater is drained out if outlets are available.

Disadvantages

1. More labour is required for levelling of the field.
2. Ridges cut down the neat cropped area.
3. Large irrigation streams are required.
4. Repair of ridges and supervision during irrigation are needed.

Ring/Ring Basin Method

Ring basin method of irrigation

Ring basin irrigation is especially suitable for fruit and vegetable crops. A basin is made around the stump of the plant at a distance of about 30 -60 cm which may be square, rectangular or circular in shape. The individual plant or plant grown in pit is irrigated by this method. The soil around the plant is soaked with irrigation water and not the entire land. Generally ring basin is used for irrigating fruit trees. The basin is connected with irrigation channels for receiving water.

Advantages

1. It is a good method for irrigating orchard plants and also for all types of soil.
2. It is an efficient method in the use of water.
3. This method economise the water considerably.

Disadvantages

1. Initial cost for preparing basin is high.
2. This method is not suitable for irrigating field crops and some orchard plants such as banana, papaya, pineapple etc.
3. All plants do not get same amount of water.

Sprinkler Irrigation System

Sprinkler irrigation system allows application of water under high pressure with the help of a pump. It releases water similar to rainfall through a small diameter nozzle placed in the pipes. Water is distributed through a system of pipes, sprayed into air and irrigates in most of the soil type due to wide range of discharge capacity.

Advantages

- Eliminates water conveyance channels, thereby reducing conveyance loss.
- Suitable in all types of soil except heavy clay.
- Water saving up to 30% - 50 %.
- Suitable for irrigation where the plant population per unit area is very high.
- Helps to increase yield.
- Reduces soil compaction.
- Mobility of system helps system operation easy.
- Suitable for undulating land.

- Saves land as no bunds required.
- Soluble fertilizers and chemicals use are possible.
- Provides frost protection & helps in alteration of micro climate.
- Reduces labour cost.

Components

Major Components of Sprinkler Irrigation System	
1	Pumping station or Header Assembly
2	By-pass valve
3	Fertilizer tank
4	Filtration system
5	Pressure gauges
6	Control valves
7	HDPE / PVC Pipes
8	QRC Pump connector
9	Sprinkler Nozzles
10	Service Saddle

Sprinkler Layout

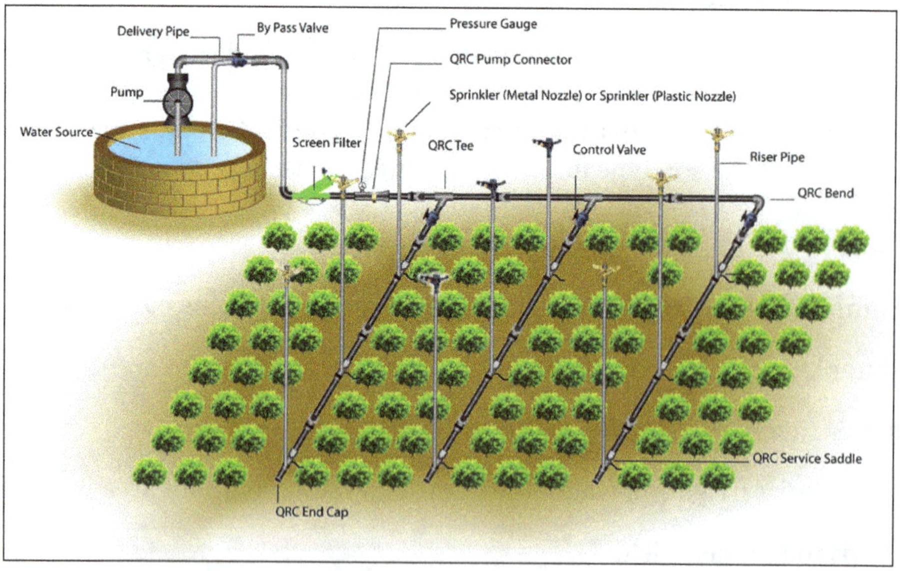

A typical sprinkler irrigation system consists of the following components:

1. Pump unit
2. Mainline and sometimes sub-mainlines

3. Laterals
4. Sprinklers.

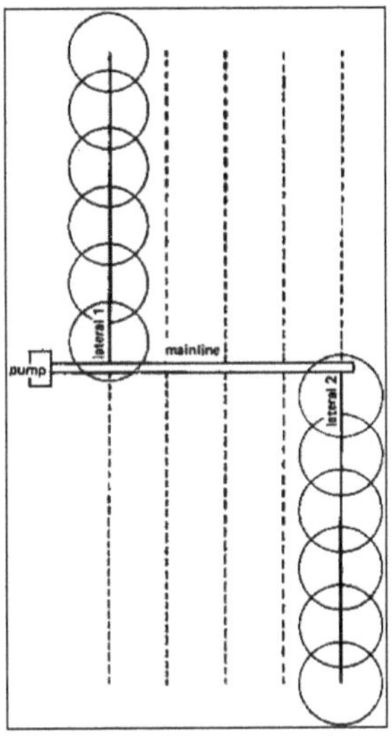

Types of Sprinklers Based on Precipitation:

Type of Sprinkler	Precipitation Rate (mm per hour)
Low Volume Sprinkler	Less than 13
Medium Volume Sprinkler	13 - 25
Large Volume Sprinkler(Raingun)	Above 25

Response of Different Crops to Sprinkler Irrigation System:

Crops	Water Saving (%)	Yield increase (%)
Chilli	33	24
Cotton	36	50
Fenugreek	29	35
Gram	69	57
Jowar	55	34
Maize	41	36
Onion	33	23
Sunflower	33	20
Wheat	35	24

Wetting Patterns in Sprinklers

The wetting pattern from a single rotary sprinkler is not very uniform. Normally the area wetted is circular. The heaviest wetting is close to the sprinkler. For good uniformity several sprinklers must be operated close together so that their patterns overlap. For good uniformity the overlap should be at least 65% of the wetted diameter. This determines the maximum spacing between sprinklers.

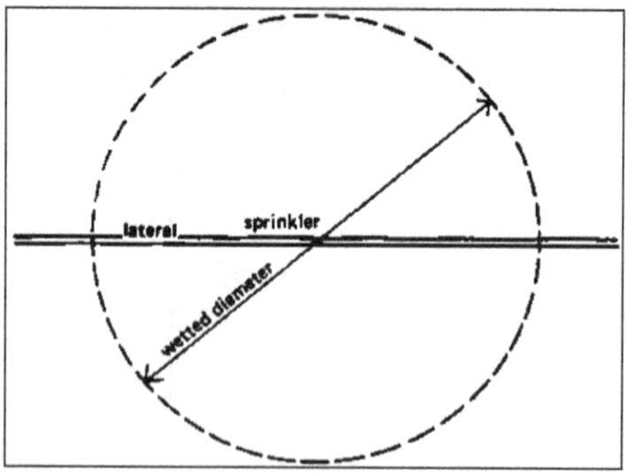

Wetting pattern for a single sprinkler (Top view)

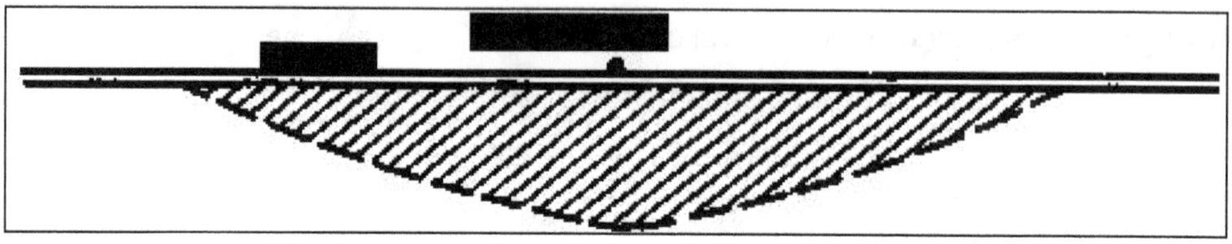

Wetting pattern for a single sprinkler (Side view)

The uniformity of sprinkler applications can be affected by wind and water pressure. Spray from sprinklers is easily blown about by even a gentle breeze and this can seriously reduce uniformity. To reduce the effects of wind the sprinklers can be positioned more closely together.

Sprinklers will only work well at the right operating pressure recommended by the manufacturer. If the pressure is above or below this then the distribution will be affected. The most common problem is when the pressure is too low. This happens when pumps and pipes wear. Friction increases and so pressure at the sprinkler reduces. The result is that the water jet does not break up and all the water tends to fall in one area towards the outside of the wetted circle. If the pressure is too high then the distribution will also be poor. A fine spray develops which falls close to the sprinkler.

- Application rate: This is the average rate at which water is sprayed onto the crops and is measured in mm/hour. The application rate depends on the size of sprinkler nozzles, the operating pressure and the distance between sprinklers. When selecting a sprinkler system it is important to make sure that the average application rate is less than the basic infiltration rate of the soil. In this way all the water applied will be readily absorbed by the soil and there should be no runoff.

- Sprinkler drop sizes: As water sprays from a sprinkler it breaks up into small drops between 0.5 and 4.0 mm in size. The small drops fall close to the sprinkler whereas the larger ones fall close to the edge of the wetted circle. Large drops can damage delicate crops and soils and so in such conditions it is best to use the smaller sprinklers.

Drip Irrigation

Drip irrigation is renowned for being a very efficient method of watering plants. For example, the average sprinkler system has an efficiency of around 75-85%. Drip irrigation, in contrast, has efficiencies in excess of 90%. Over time this difference in water delivery efficiency will make a considerable difference in crop production and the company's bottom line. In areas where water is short supply, such as the desert regions of the US, drip irrigation has, unsurprisingly, become the preferred method of irrigation. Drip irrigation systems are relatively inexpensive and easy to install, simple to design, and help maximize plant health due to the reduced moisture levels on fields.

This form of irrigation, sometimes called trickle irrigation, applies water directly and slowly to the soil. The technique's efficiency is provided by two major factors. The first is that water is absorbed by the soil for access to the plant roots rather than running off or evaporating. Secondly, the water is only supplied to those areas of the field that actually need the water, i.e. the plant roots. Most drip irrigation systems are simple to design which minimizes design errors and installation flaws.

Importance of Irrigation

Irrigation is one of the oldest technologies mankind developed. It is used extensively across the world.

Drip irrigation offers a great solution to this potential problem. Historic practices such as centre-pivot irrigation are not sustainable in the long term. They consume large amounts of water and are potentially damaging to the "health" of the soil. Drip irrigation allows the user to better control the amount of water plants receive, rather than blanket watering the area. Eutrophication is massively decreased by drip irrigation as fertilizers are not carried away by water runoff into watercourses.

Working Principle of Drip Irrigation

In effect, drip irrigation places small drip emitters in close proximity to the crops' root systems. This provides a much-improved efficiency and makes the system much more controllable compared to other methods. The emitters release water in a slow and steady fashion. The emitters are very small, about the size of a US quarter and are arranged in an array in the ground. These emitters are directly connected to the water source by feeder hoses. Another setup is to have the emitters built into the feeder hose instead of rows of independent emitters. This is called a trickle hose.

Drip irrigation was tested in a primitive form in the 1920's but the modern technology as we know it was properly developed by Simcha in the 1930's in Israel. Its discovery seems to have been something of an accident. Blass, whilst spending some time in the desert regions of Southern Israel noticed something strange. He noticed that one tree near his location was performing much better than all other vegetation nearby.

When Blass went for a closer look, he noticed that a water pipe near the tree had a small leak supplying its root system with its very own regular slow supply. This accidental discovery led Blass to embark on a journey of trial and error testing various materials and water pressures for an ideal solution. It wasn't until the 1950's with modern plastics that Blass could take his technology to the next level. In the 1960's Blass was able to finalize the technology and patent the design.

Components of Drip Irrigation System

Drip irrigation systems are pretty simple setups but do consist of several constituent parts. A typical simple system will consist of the following components.

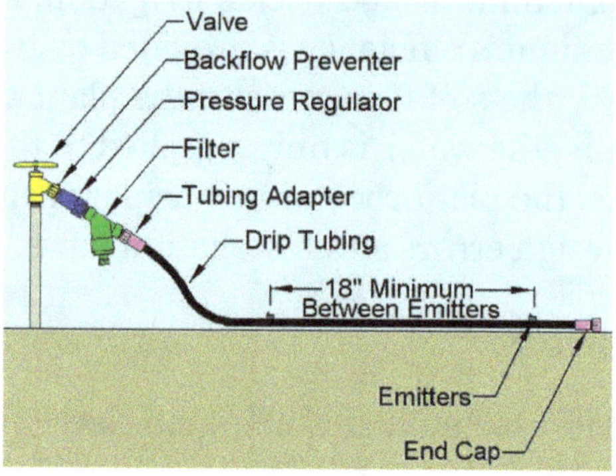

Simplified drip irrigation system

Valves

The role of valves in the drip irrigation system is very simple. They turn the water flow on or off. Valves come in various "flavors". Isolation valves are manually operated for systems that require an infrequent shut-off of the water. These valves are typically sited close to the water supply to enable isolating the system for repairs or off season. These can be installed anywhere in the system to enable isolation of segments of the system for localized repairs but this is usually only employed in larger systems.

Control valves are valves that turn the water on and off to individual "circuits" or areas of the yard that are perhaps irrigated separately from one another. These can be automatic (using solenoids) or manually operated. Depending on the system design there can be just one or many installed. For instance, you may have one control valve that controls water supply to the emitters in a vegetable garden. Yet another may be present that controls water supply to shrubs or hanging pots around the house and patio.

Drip irrigation system

Backflow Preventer

This is a piece of kit employed within the system to prevent, hence the name, dirt, bacteria, other contaminants from being sucked back up into the drinking water supply for the drip system. This device is essential for all drip irrigation systems.

Backflow preventers are essential because drip emitters rest directly on the soil and are potential very susceptible to water contamination from soil born diseases etc.

Pressure Regulators and Pressure Reducing Valves

These devices, as the name implies, reduce the pressure of water flowing through the system and keep it at a constant level. Pressure reducing valves and pressure regulators are in this instance synonyms and essentially the same thing.

Drip irrigation systems, on the whole, perform best at lower water pressures than typical water supply systems. These devices also enable a constant system pressure even if the supply pressure fluctuates periodically, which is nice. Designers need to be conscious of areas with low water pressure as these devices will clearly further reduce the system pressure.

Usually, two types of pressure regulators are employed in drip irrigation systems. Non-adjustable ones with pre-set outlet pressures and user adjustable types. In general, small homeowner system will use non-adjustable valves if they have less than 3 control valves. Non-adjustable regulators must be installed after the control valve and in cases where multiple control valves are present, pressure regulators are needed for each one. Accidental installation before the control valves can cause pressure surges that will damage the system.

Adjustable pressure regulators, on the other hand, can be installed before or after the control valves. In large systems, you can install a single or fewer adjustable pressure regulators in the main supply line before control valves to save on costs.

Filters

The filter is used to, well, filter the water. Drip emitters have very small openings which are easily clogged up so employing filters earlier in the system is essential for increasing the life span of the irrigation system.

Emitters

Now we come to the "guts" of the drip irrigation system. The emitters are responsible for directly controlling the rate of water supply to the soil. Emitters are usually small plastic devices that either screw or snap onto the drip tube or pipe. In trickle pipe systems they are pre-assembled and part of the pipe assembly.

As a general rule of thumb 1 or 2 emitters are usually required per plant. This does, of course, depend entirely on the size of the plant in question. Trees or shrubs will clearly need more than a small plant. Use of multiple emitters also provides the system with backups in case of a blockage in one or more of the emitters. The more emitters present the wider the area of irrigation and hence the increased growth of roots for healthier crops and plants.

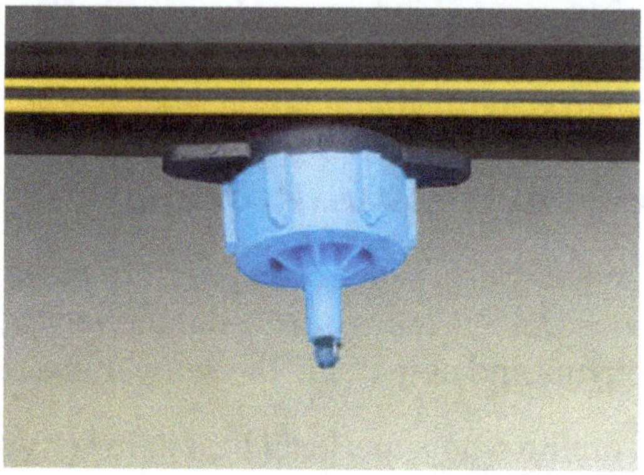

Mainline and Lateral/Sub-main Pipes

This pipe is the main connection between the water supply to the control valves of the drip irrigation system. It can be made of galvanized steel, copper, PVC or heavy wall Polyethylene. Each type has the inherent limitations and strengths. PVC, for example, is easily damaged by sunlight and is usually buried or protected. Polyethylene has a low burst pressure and is usually only used where water pressures are lower than 50 PSI.

Lateral/sub-main pipes are located between the control valve and drip emitter assemblies. These can again be made of PVC, PEX or Polyethylene. As these are generally placed downstream of the pressure regulator high-pressure ratings are not essential.

Drip Tube or Hose

This is a special type of tube common in most drip systems. They tend to be laid on the ground surface between plants. Emitters are generally installed to these tubes. Drip tubes tend to be made of thin-walled polyethylene and consequently have a much lower pressure rating than other parts of the system. It is generally recommended that these stay above ground as they can commonly be nibbled on by pesky local rodents! In large commercial setups, these tubes are usually "hard-piped" in these systems and the emitters are installed directly onto the laterals.

Air Vent

Air vents are installed in systems that are turned off at any time. They prevent air from being sucked up into the emitters. As the water pressure falls away, air can be sucked back through the emitters and entrain dirt or soil into them. Clearly not desirable. The presence of an air vent mitigates this problem by drawing in air through it rather than the more delicate emitter openings.

Advantages and Disadvantages

The advantages of drip irrigation are:

- Fertilizer and nutrient loss is minimized due to a localized application and reduced the leaching.
- Water application efficiency is high if managed correctly.
- Field leveling is not necessary.
- Fields with irregular shapes are easily accommodated.
- Recycled non-potable water can be safely used.
- Moisture within the root zone can be maintained at field capacity.
- Soil type plays a less important role in the frequency of irrigation.
- Soil erosion is lessened.
- Weed growth is lessened.
- Water distribution is highly uniform, controlled by the output of each nozzle.
- Labour cost is less than other irrigation methods.
- Variation in supply can be regulated by regulating the valves and drippers.
- Fertigation can easily be included with minimal waste of fertilizers.
- Foliage remains dry, reducing the risk of disease.
- Usually operated at lower pressure than other types of pressurized irrigation, reducing energy costs.

Drip irrigation and spare drip irrigation tubes in banana farm

The disadvantages of drip irrigation are:

- Initial cost can be more than overhead systems.
- The sun can affect the tubes used for drip irrigation, shortening their lifespan.
- The risks of degrading plastic affecting the soil content and food crops. With many types of plastic, when the sun degrades the plastic, causing it to become brittle, the estrogenic chemicals (that is, chemicals replicating female hormones) which would cause the plastic to retain flexibility have been released into the surrounding environment.
- If the water is not properly filtered and the equipment not properly maintained, it can result in clogging or bioclogging.
- For subsurface drip the irrigator cannot see the water that is applied. This may lead to the farmer either applying too much water (low efficiency) or an insufficient amount of water, this is particularly common for those with less experience with drip irrigation.
- Drip irrigation might be unsatisfactory if herbicides or top dressed fertilizers need sprinkler irrigation for activation.
- Drip tape causes extra cleanup costs after harvest. Users need to plan for drip tape winding, disposal, recycling or reuse.
- Waste of water, time and harvest, if not installed properly. These systems require careful study of all the relevant factors like land topography, soil, water, crop and agro-climatic conditions, and suitability of drip irrigation system and its components.
- In lighter soils subsurface drip may be unable to wet the soil surface for germination. Requires careful consideration of the installation depth.

- Most drip systems are designed for high efficiency, meaning little or no leaching fraction. Without sufficient leaching, salts applied with the irrigation water may build up in the root zone, usually at the edge of the wetting pattern. On the other hand, drip irrigation avoids the high capillary potential of traditional surface-applied irrigation, which can draw salt deposits up from deposits.
- The PVC pipes often suffer from rodent damage, requiring replacement of the entire tube and increasing expenses.
- Drip irrigation systems cannot be used for damage control by night frosts (like in the case of sprinkler irrigation systems).

Pot irrigation by On-line drippers

Pressure compensated integral dripper on soilless without growing channels

Drip Tape

Drip tape is a type of thin-walled dripperline used in drip irrigation. The first drip tape was known as “Dew Hose”.

Drip tape duct tape is made of polyethylene and is sold flat on reels. The wall thickness typically ranges from 4 to 25 mils (0.1–0.6 mm). Thicker walled tapes are commonly used for permanent subsurface drip irrigation and thinner walled tapes for temporary throw-away type systems in high-value crops.

Water exits from tape through emitters or drippers. The typical emitter spacing ranges from 6 to 24 inches (150–600 mm). In some products, the emitters are manufactured simultaneously with the tape and are actually formed as part of the product itself. In others, the emitters are manufactured separately and installed at the time of production.

Some product is not a tape, but a thin-walled dripperline, but in popular parlance, both types of products are called tapes. Typical tape diameters are 5/8”, 7/8”, and 1-3/8”, with the larger diameters more commonly used on permanent installations with longer runs.

Drip tape is a recyclable material and can be recycled into viable plastic resins for reuse in the plastics manufacturing industry.

Uses

Irrigation dripper

Drip irrigation is used in farms, commercial greenhouses, and residential gardens. Drip irrigation is adopted extensively in areas of acute water scarcity and especially for crops and trees such as coconuts, containerized landscape trees, grapes, bananas, ber, eggplant, citrus, strawberries, sugarcane, cotton, maize, and tomatoes.

Drip irrigation for garden available in drip kits are increasingly popular for the homeowner and consist of a timer, hose and emitter. Hoses that are 4 mm in diameter are used to irrigate flower pots.

Subsurface Drip Irrigation

Subsurface drip irrigation is a variation from conventional surface drip irrigation techniques. By minimising evaporation it uses water more efficiently than surface irrigation. The depth that the laterals (also used in conventional drip irrigation) are buried at depends mostly on the tillage practices and the crop to be irrigated. It is a costly system that needs expert design and maintenance. But it is also a very effective method that transports the water directly to the roots and is especially appropriate for arid and windy locations. Before entering the distribution pipes the water is filtered to minimise the risk of clogging. Still, the system must be designed so that all pipes can be flushed periodically. As it requires high maintenance and expert design, subsurface drip irrigation systems are more suitable for medium and large-scale production.

Advantages

- High degree of control over water application with the potential for high uniformity of application.
- Evaporation is reduced.
- The amount of water can be fine-tuned. This avoids water loss caused by run off or evaporation.

- Frequent irrigation allows for optimum soil moisture content in the root zone.
- Great performance in windy and arid locations.
- If pre-treated wastewater is used for irrigation, the risk of direct contact with crops and labourers is reduced in Precipitation, Freshwater, Fertigation Water.

Disadvantages

- Risk of clogging.
- When saline water is used, salts accumulate at the wetting front.
- Emitter can be damaged or blocked by root hairs.
- Bacterial slimes and algae growing on the interior walls of the laterals and emitters combined with clay particles in the water can block the emitters.
- Suspended organic matter and clay particles can damage the system.
- A lot of repair work is caused by rodents chewing the tubes.
- Heavy machinery can damage the laterals.

Food Products

A field irrigated by a subsurface drip system.

Subsurface drip is a highly efficient irrigation system that uses buried drip tubes or drip tape to meet crop water needs. Since the water is applied below the soil surface (as opposed to surface irrigation or traditional drip irrigation), the effects of surface infiltration, such as crusting, saturated condition of ponding water, and water losses via evaporation and surface runoff (including soil erosion) are eliminated. With an appropriately sized and well-maintained subsurface drip irrigation system, water application is highly efficient and uniform.

Wetting occurs around the tube and water moves out in all directions. Moreover, water is applied directly to the root zone of the crop as opposed to the soil surface where most weed seeds hibernate. As a result, germination of annual weed is reduced. This lowers the pressure on valuable crops. Furthermore, some crops may benefit from the additional heat provided by dry surface conditions, and produce more biomass. When managed properly the application of fertiliser can be optimised. Fields can still be worked when irrigation systems are installed.

Basic Design Principles

A subsurface drip irrigation system has a similar design as a common drip irrigation system. A typical system layout consists of a settling pond (where possible), pumping unit, pressure relief valve, check valve or back flow prevention valves, hydrocyclone separator (if a settling pond is not feasible), chemical/fertiliser injection unit, filtration unit equipped with back flush valves, pressure regulators, air vent valves and PVC pipes delivering the water to the crop. The piping is 10 to 60 cm below the ground, depending on crop and soil (capillary attraction). As a water source, treated greywater or even blackwater is possible, with the risk of clogging being greater if the influent flow has not properly settled. Therefore, treatment of the water (e.g. a non-planted filter system, constructed wetlands (horizontal flow or vertical flow) or at least a septic tank) before the settling pond is necessary.

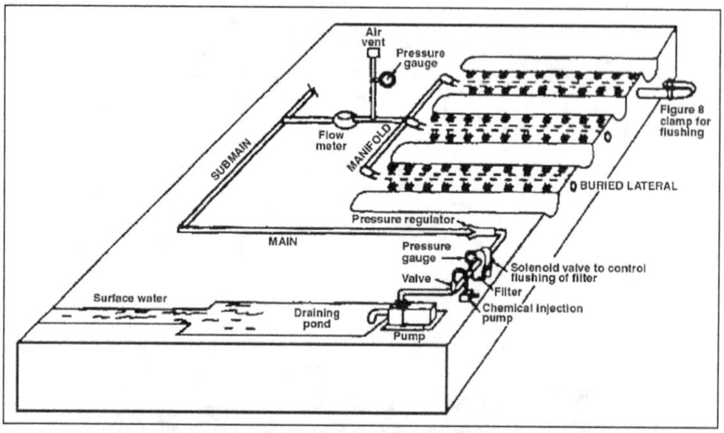

A typical subsurface drip irrigation field layout.

Subsurface drip irrigation is generally a high-tech, automatically operated technology applied to medium and large scale production. However, several low-cost and simple methods of subsurface (drip) irrigation like pitcher or bottle irrigation exist that are equally effective for small-scale farming. There are several subsurface techniques used for secondary wastewater treatment such as leach field or evapotranspiration bed that also provide uncontrolled irrigation to fields.

Operation and Maintenance

The performance and life of any system depends on how well it is designed, operated and maintained. It should either be automatically controlled or regularly inspected. Repairing the buried pipes is difficult, cumbersome and time-consuming. To prevent rodents from chewing the pipes, precautionary measures should be undertaken. The mechanical components such as pumps, valves and filters need to be maintained as well as checked and cleaned regularly.

Drip irrigation requires energy to deliver the water.

The filter system of a large-scale subsurface drip irrigation facility is complex and its proper functioning is crucial to the whole system. The maintenance of the three main filters of the system (this contains centrifugal separators, screen and disk filters as well as sand media filters) should be carried out carefully (e.g. backwashing of sand media filter). Systems must be designed so that mainlines, sub-mains, manifolds and laterals can all be flushed. Mainlines, sub-mains and manifolds are flushed with a valve installed at the very end of each line. Lateral lines can be flushed manually or automatically. It is important to flush the lines at least every 2 weeks during the growing season.

Drainage System

In Agriculture, Drainage is the in agriculture, the artificial removal of water from land; drainage is employed in the reclamation of wetlands, in the prevention of erosion, and as a concomitant of irrigation in the agriculture of arid regions.

Drainage is an ancient practice, but apparently until recent times it was regarded as less important than irrigation. The first drains were most likely ditches for channelling floodwaters back to the rivers. The addition of linings of less porous materials greatly improved drainage efficiency. The most significant 20th-century development in drainage technology was the application of land-grading techniques to facilitate uniform runoff.

Land may be smoothed with proper slopes and ditches so as to remove excess water before it enters the soil and thus prevent erosion, leaching of nutrients, and standing pools of water on the surface, and to permit early spring planting. If carefully planned, this smoothing also can prepare the land for surface irrigation, thus serving two purposes by one earth-moving operation. After excess water enters the soil, its removal is an expensive and specialized undertaking that is not directly connected with irrigation, although it sometimes may be necessary for irrigated land. Modern drainage systems may be divided into two categories, surface and subsurface. The typical surface system consists of field drains, field ditches, a main collection ditch, and an outlet. As the term implies, a surface system is designed to remove water that collects on top of the soil. Surface drainage is especially important for soils that absorb water slowly. The field drains vary in configuration according to topography, parallel drains being indicated for uniform surfaces and site-specific ones for areas of uneven accumulation.

Subsurface drainage systems consist of small conduits, a submain, a main, and an outlet. The conduits, equivalent to the field drains in a surface system, collect the water in the soil and drain it into the larger arteries.

Factors determining the most efficient drainage system design for a particular property include soil type, land configuration, amount and pattern of rainfall, and types of crops to be grown. Soils of high sand or silt content are generally suited to subsurface drainage, while soils of high clay content generally require surface systems.

Methods of Drainage

Surface Drainage

Surface drainage is very useful in removing excess water from land in a controlled manner and as quickly as possible, to a artificial drainage system or a natural watercourse. This should be done with no damage to the environment.

Wet soil problems on dairy pastures are usually due to both excess surface and sub surface water. It is important to realise that surface drains will only remove surface water, and in most soil types, are almost useless for draining the soil profile.

Never-the-less surface drains are essential and very useful tool to manage wet soils when used in the right location for the right reasons. Typical applications for using surface drains include:

- Draining large quantities of surface water off land quickly.
- Intercepting water flowing down from upper slopes to lower lying land.

- Intercepting run off water from other sources (eg. road side drains, neighbouring properties, dam overflows, etc.
- Acting as collector system for sub surface (mole or tile drains) drainage systems.

Types of Surface Drainage Systems

Ditches or Open Drains

These vary in size and length and can be simply categorised as follows:

- Shallow surface drains (up to ~0.3m deep) formed by hand shovel or "spinner cuts" are most useful for removing ponded water from shallow depressions to larger drains or streams. They are generally not suitable for draining large flat areas. They tend to be temporary in nature since they are often pugged by animals, silt up quickly, over grow with less desirable species, wrecked by machinery in wet weather and so on. They are very inconvenient for feedout or passenger vehicles and fodder conservation equipment.
- Medium sized drains (0.3m to ~ 1m deep) made by excavators or specific drainage machines are usually "V" shaped with a flat bottom with enough slope (gradient) to ensure that the water is removed quickly but not cause scouring of the drain walls or bottom. This is usually not a problem because they are used on the flatter areas. The velocity of water flow should be slow on the sandy and sandy clay loam soils and can be quicker on the clays and clay loams.

The slope or batter of the drain banks should be such that the banks do not collapse when wet. Each soil type has its own natural slope that will not collapse and machine formed banks should be similar or even less steep. Clays may have a batter of 1:1, silty and clay loams 1:1 to 1:1.5, and sandy loams 1:1.5 to 1:2, while looser sandier soils should be 1:2 to 1:3, shows what happens to drains with insufficient batter angle.

The size of open drains depends on the amount of water to be removed or intercepted and formulae are available to calculate the required design parameters.

The time to construct open drains is when the topsoil and sub soil is soft enough to dig but water presence is not a problem, usually in late spring early summer and autumn.

Caution:

- Be very wary of Constructing open Drains in Dispersive soil as they are Highly prone to Erosion.
- Large open drains or canals (several meters deep and wide) made by dragline excavators, scrapers or bulldozers carry very large volumes of water and generally not applicable to most dairy farms apart from possibly delivering irrigation water or its drainage so will not be covered here.

Disadvantages of Open Drains

- They need regular cleaning (sediment and weeds).
- Can be dangerous to animals, machinery, and people (if unfenced).
- If fenced, cleaning is more difficult.
- Fencing adds extra cost and maintenance.
- Tend to become deeper and wider if not maintained by professional machinery operators.
- Not suitable for sub-surface drainage.
- Loss of ground.

Medium sized "V " shaped open drain

Grassed Waterways

Grassed waterways are usually shallow, varying in width from narrow to meters wide, but are constructed such that they are often grazed as part of the paddock. They are sometimes used to bring drain outflows down slopes to prevent erosion without considerable expense.

Open drain with incorrect bank matter

Their steepness should be less than 1:4 (vertical : horizontal). A dense and unbroken grass sward is essential and should be established before it is used to discharge water.

Avoid grazing the waterway while it is wet or draining water, particularly if the pasture is not well established.

The amount of water to be removed and the steepness of the slope will determine the size and shape of the waterway. A good rule of thumb is that the steeper the fall the wider the waterway. The bottom of the waterway should be as near to horizontal in cross section as possible so that the water spreads out evenly over the base.

The waterway must not be constructed in soils prone to erosion. Water passing down slopes which have clay sub soils should avoid saturation of large areas of this clay as it may result in landslips. Water passing down slopes should do so quickly in as narrow a channel as possible but avoiding the possibility of erosion. Dense pasture swards assist greatly in soil stabilisation.

Humps and Hollows (Bedding)

Hump and hollowing is the practice of forming (usually while renovating pastures) the ground surface into parallel convex (humps) surfaces separated by hollows. The humped shape sheds excess moisture relatively quickly while the hollows act as shallow surface drains.

Humps and hollows are useful in areas or on soil types that are not suitable for tile or mole drainage. They are also useful where the lack of suitable outfalls prohibit the use of tile drains, usually due to insufficient depth or fall.

The humps and hollows system comprises lateral surface drains which discharge into headland drains which in turn discharge via short open drains or shallow pipes into natural watercourses or open drains. If possible these areas should not be grazed while the drains are still wet because they will become pugged. If this does occur then end of season maintenance must ensure that all the drains are reshaped or cleared to allow quick drainage.

The size of this drain type depends on the amount of water to be removed or intercepted and formulae are available to calculate the required design parameters.

Dimensions

Humps and hollows in newly sown pasture

Humps may vary from 10 – 20 m spacings but the wider the hump, the more soil that must be moved, and the water run off may be substantially slower. Humps wider than this may not drain at all in their centres. Gippsland experiences suggest a spacing of approximately 15 m width and approximately 0.4 – 0.5 m depth. For drains discharging in one direction, lengths should not exceed about 150 m, but can be longer if discharging in two directions.

Disadvantages

Humps and hollows make it more difficult for machinery activity such as fodder conservation. The humps may be overgrazed while the hollows may become pugged. The depth of topsoil must be enough to allow the humps and hollows to be formed without exposing the poorer and nutrient deficient subsoil. Alternatively more soil will be required to be moved to ensure a covering of topsoil in the hollows for resowing pasture. If the edge of the hump is too steep cattle trampling could cause soil movement into the hollows and possibly into watercourses.

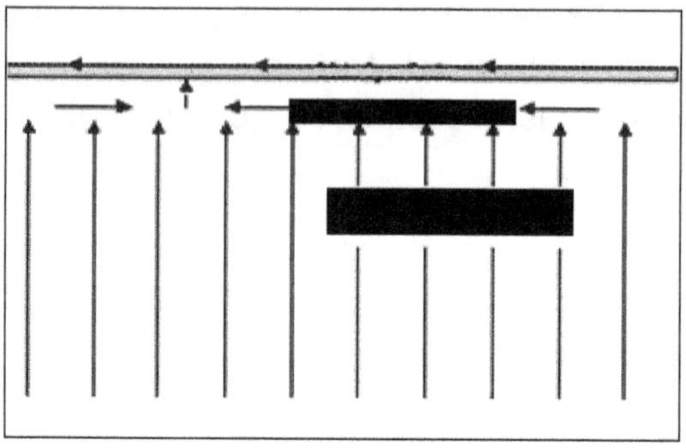

A typical design for hump and hollow drainage system

Levees or Graded Banks

These are surface channels, often formed on sloping land, in such a way that the earth removed forms a bank or levee on the downslope of the channel. Where used on slopes, the main purpose of graded banks is to prevent surface runoff from building up volumes and velocity that may cause erosion.

The banks must follow the contour of the slope with a gradual decline so that water flows constantly and does not build up in depressions, thus rendering land below susceptible to "landslips ". The outfalls of the graded banks may possibly lead to dams for water collection for irrigation, stock water, etc. Bank spacing will depend on slope gradient, amount of water to be collected, outfall locations, slope topography and amount of rainfall.

Dimensions

Spacings of about 30 - 50 m for slopes with gradients of 50 - 120 have proven acceptable in most areas. Each bank should service a total area of up to about 3 hectares, with length not exceeding about 400m.

The channel will be at risk to erosion immediately after construction until a grass cover is established. Care must be taken on slopes that may have shallow topsoil over dispersive subsoil, as severe erosion is possible in these soil types.

Levee banks may be used to confine large flows of water to certain areas and are often used to control floodwaters along streams. However approval for use of levee banks in this situation is required from the local or regional water authorities.

Laser Levelling (Grading)

Laser levelling is a relatively new practise used on irrigation farms to even out and speed up the water flow off flood irrigation bays. Although not common laser levelling could be used on the "flats " areas in high rainfall zones to ensure water moves off quickly and does not remain in depressions.

Subsurface Drainage

Subsurface Drainage Is Concerned with removing water that percolates through or is contained in the underlying subgrade. This water, typically the result of a high water table or exceptionally wet weather, can accumulate under the pavement structure by two chief means:

- Gravity flow: Water from surrounding areas can be absorbed by the soil then flow by gravity to areas underneath the pavement structure. In pavement with high air voids (above 8 – 9 percent), water can percolate down through the pavement structure itself.
- Capillary rise: Capillary rise is the rise in a liquid above the level of zero pressure due to a net upward force produced by the attraction of the water molecules to a solid surface (e.g., soil). Capillary rise can be substantial, up to 6 m (20 ft.) or more. In general, the smaller the soil grain size, the greater the potential for capillary rise. Often, capillary rise is a problem in areas of high groundwater tables.

Most pavements have performed adequately without considering these effects. However, Hot Mix Asphalt pavements can fail because of subgrade support deterioration as a result of excessive moisture or other water-related problems. While the best solution is usually to prevent water infiltration with surface drainage measures, subsurface drainage can be useful, however it needs to be done judiciously, because it may be somewhat akin to treating the symptom rather than the problem. Subsurface drainage consists of three basic elements:

- A permeable base to provide for rapid removal of water which enters the pavement structure.asphalt treated permeable base layers may strip and become clogged with fines thus weakening the overall pavement structure.

- A method of conveying the removed water away from the pavement structure. At the least, this may consist of a base sloped towards a drainage ditch. At the most, this may consist of a pipe collector system.
- A filter layer (such as a geotextile, graded aggregate layer or Hot Mix Asphalt) to prevent the migration of fines into the permeable base from the subgrade, subbase or shoulder base material. Excess fines in the permeable base will clog its drainage routes and render it ineffective. Depending upon the subgrade and pavement structure a filter layer may not be used.

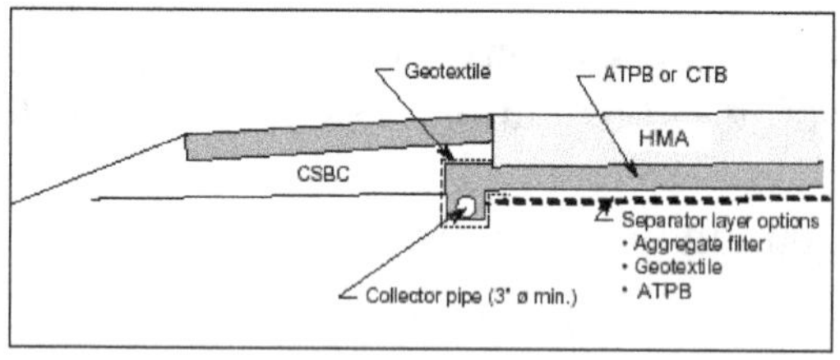

Hot Mix Asphalt pavement subsurface drainage

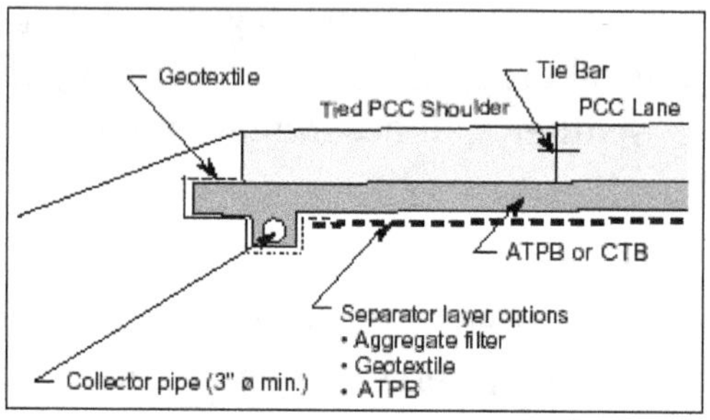

PCC pavement subsurface drainage with PCC tied shoulder

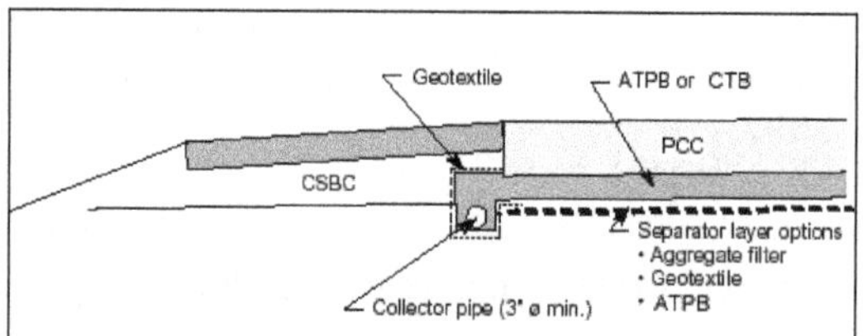

PCC pavement subsurface drainage with Hot Mix Asphalt shoulder

Benefits of a Subsurface Drainage System

Costing of drainage systems has shown that both increased pasture utilisation and extra pasture growth from drainage can be very competitive with other feed sources. However to take advantage of this extra growth, more cows, and farm improvements (including a feed pad) may be necessary.

However, subsurface drainage provides many intangible benefits such as improvement in soil health due to the increased aeration of the soil, increased responses to fertiliser use, reduced mineral imbalances in the soil and sometimes in the plants as well, are long term sustainability benefits. Salinity can be prevented or greatly reduced if in its early stages.

Drained soils enables fodder conservation of silage/hay to occur several weeks earlier than on undrained soils. Therefore the resultant fodder has the potential to be of much higher quality producing improved animal production.

Drainage, by reducing pugging and creating favourable soil environments, results in the retention of the improved pasture species versus the influx of plants suited to waterlogged conditions such as rushes, fog grass, glyceria species, water couch, etc.

Most importantly, reduced stress in managing stock and pastures during the wetter months of winter or spring has very large benefits for the farm operator.

Types of Subsurface Drainage Systems

There are four main types of subsurface drainage systems. These are:

- Corrugated and PVC slotted subsurface pipes.
- Mole drainage including.
- Mole drains Mole drains over collector pipe systems Gravel mole drains.
- Interceptor drains.
- Ground water pumps.

Subsurface Pipes

Subsurface pipes can be used to drain heavy (clay) poorly drained soils successfully but their spacing would need to be so close together that they are uneconomic in extensive farmland systems.

Subsurface pipe drainage was referred to as 'tile drainage' in the past due to the use of short clay pipes. Clay was expensive and difficult to lay and has now been replaced by slotted PVC or flexible corrugated plastic pipes of variable diameters.

Drainage trencher installing subsurface pipe drains and depositing permeable backfill on top

Specifically designed drainage trenchers,usually fitted with laser guidance equipment, dig the trench, lay the slotted pipe and place permeable backfill into the trench on top of the laid pipe. This backfill is delivered by trucks or trailers fitted with conveyor belts which feed the backfill into the hopper. The forward speed, hopper channel opening size, material size, etc. determine the depth and amount of material laid on top of the pipe.

In very permeable soils, very little backfill is needed but in less permeable soils, or where moles are to be pulled through above a pipe, the backfill depth reaches to near the ground surface.

Mole Drainage

Mole drainage can be classified as mole drains, mole drains over a collector pipe system or gravel mole drains. The suitability of each type will depend on the clay content and type, sand and stone in the profile, gradient and outfall location. The action of the mole plough forms a mole channel in the area of the soil profile with a specific clay content. The plough also cracks the soil profile immediately above the mole channel allowing water to flow into it.

Mole Drains

Mole drains are used in heavy soils where clay subsoil near moling depth (400 to 600cm) prevents downward movement of ground water.

Mole plough

The success and longevity of mole drains is dependent on soils having a high clay content so that once a mole channel is formed, it will maintain the channel for many years. Mole drains are not suited to soils with clay types which have dispersive or slaking characteristics. Mole drains are also not suited to permeable soils due to their high sand and loam contents.

A mole plough is used to form mole drains. Simply, a mole plough contains a leg (or blade) to which a torpedo (or foot) is attached to its bottom. Sometimes a plug (or expander) and having a slightly larger diameter, is attached to the rear of the torpedo, and ensures the mole channel is left with the correct shape.

Mole Drains over Collector Pipe Systems

This system is used in soils where it is not possible to form moles that reach the outfall. This includes the presence of stones, sandy pockets, uneven surfaces or excessive distances to the outfall.

In heavy soils where mole drains would need to be very long (over 80 m) before they reach an outfall, installing subsurface pipes at approximately 60 to 100 m, over which mole drains are pulled, can be very successful. Washed sand or small diameter gravel is backfilled into the pipe trench to near the ground surface at installation.

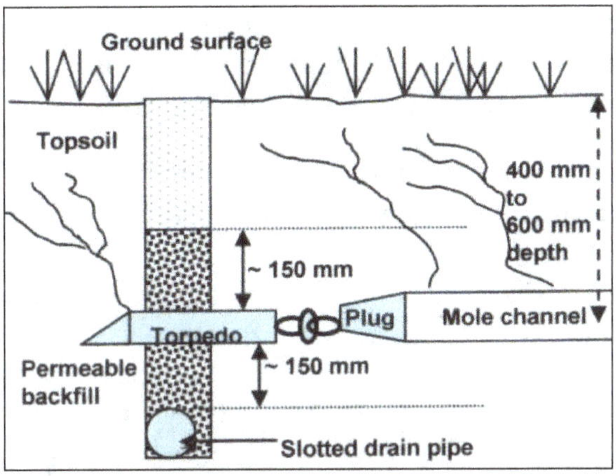

Mole drain over collector pipe System

Mole drains are then installed at or close to a right angle to the direction of the pipes. Excess ground water flows into and along the mole drains, then drains into the porous backfill above the pipes, and then is quickly removed to outfalls via the subsurface collector pipes.

Gravel Mole Drains

Gravel mole drains are best suited to soils and situations where subsurface pipes are unsuitable, where mole drains have a very short life span, or in slaking soils so the mole channel will maintain its shape at or soon after moling. A gravel mole drain is an unlined channel and leg slot filled with small diameter gravel or washed sand. Unfortunately there are very few gravel mole drainage machines available in Australia. Also gravel mole drains are expensive due to the amount of backfill and the close spacing required. However, they do offer an alternative in some "difficult to drain" situations. They may be useful in slaking and dispersive soil types but expert opinion should be sought if considering their use in these situations.

Gravel mole drain (slot filled with gravel)

Interceptor Drains

These drains are installed at the base of slopes at the change of gradient, usually where a steeper slope meets the flats to intercept the downhill flow of subsurface water. Often the soil type on the slope is more permeable than those of the flats and this forces the water to come to the surface, usually at the change of slope.

Interceptor drains can also be installed below springs and spring lines to intercept spring water. Grazing animals severely pug the areas surrounding springs and damage is usually more concentrated down slope. This affected area increases over time as the 'soak' area spreads outward and down slope. Drainage reduces stock damage, or pugging as the soil maintains its strength and so, structure.

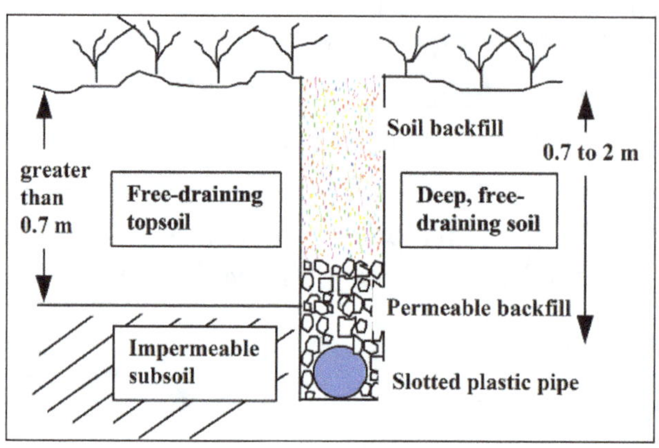

Interceptor drain

Ground Water Pumps

These remove water from aquifers so that the water table will fall or be maintained at a suitable level below the ground surface.

The act of pumping causes a drawdown of ground water leading out from the pump's location with its effect being much less at depth. The extent of effect will depend on aquifer depth, soil type, height of water table, etc.

Water Quality for Crop Production

Irrigation water quality is a critical aspect of greenhouse crop production. There are many factors which determine water quality. Among the most important are alkalinity, pH and soluble salts. But there are several other factors to consider, such as whether hard water salts such as calcium and magnesium or heavy metals that can clog irrigation systems or individual toxic ions are present. In order to determine this, water must be tested at a laboratory that is equipped to test water for agricultural irrigation purposes.

Poor quality water can be responsible for slow growth, poor aesthetic quality of the crop and, in some cases, can result in the gradual death of the plants. High soluble salts can directly injure roots, interfering with water and nutrient uptake. Salts can accumulate in plant leaf margins, causing burning of the edges. Water with high alkalinity can adversely affect the pH of the growing medium, interfering with nutrient uptake and causing nutrient deficiencies which compromise plant health.

Reclaimed water, runoff water, or recycled water may require reconditioning before use for irrigation since disease organisms, soluble salts and traces of organic chemicals may be present.

Filters

Suspended solids need to be removed from water to prevent clogging of piping, valves, nozzles and emitters in an irrigation system. Suspended solids include sand, soil, leaves, organic matter, algae and weeds. Ground water, although usually clean, may contain fine particles of sand or other particulates. All of these can be removed through filtration.

Before selecting a filter, a water analysis should be done. The type and quantity of solids should be determined, taking in consideration seasonal changes such as algae growth or spring runoff. To determine the type of filter, consider the flow rate needed to supply the irrigation system and the level of filtration needed. Screen or disk filters work well for most applications. A 200 mesh filter is usually recommended for micro-irrigation. The filter should be sized so that the flow rate is large enough to handle the peak demand.

Maintenance of a filter is important. Installing pressure gauges on both sides of the filter will indicate when it is becoming clogged. When the pressure variation between the two gauges exceeds about 10% the filter should be cleaned.

PH and Alkalinity

Alkalinity and pH are two important factors in determining the suitability of water for irrigating plants. pH is a measure of the concentration of hydrogen ions (H^+) in water or other liquids. In general, water for irrigation should have a pH between 5.0 and 7.0. Water with pH below 7.0 is termed "acidic" and water with pH above 7.0 is termed "basic"; pH 7.0 is "neutral". Sometimes the term "alkaline" is used instead of "basic" and often "alkaline" is confused with "alkalinity".

Alkalinity is a measure of the water's ability to neutralize acidity. An alkalinity test measures the level of bicarbonates, carbonates, and hydroxides in water. These compoounds get into the water from the geologic materials of the aquifer from which the water is drawn, such as limestone and dolomite. Test results are generally expressed as "ppm of calcium carbonate ($CaCO_3$)". The desirable range for irrigation water is 0 to 100 ppm calcium carbonate. Levels between 30 and 60 ppm are considered optimum for most plants.

Irrigation water tests should always include both pH and alkalinity tests. A pH test by itself is not an indication of alkalinity. Water with high alkalinity (i.e., high levels of bicarbonates or carbonates) often has a pH value of 7 or above, but water with high pH does not always have high alkalinity. This is important because high alkalinity, not pH, exerts the most significant effects on growing medium fertility and plant nutrition.

Potential Adverse Effects on Nutrition

In most cases irrigating with water having a "high pH" causes no problems as long as the alkalinity is low. High pH water has little effect on growing medium pH because it has little ability to neutralize acidity. This situation is typical for many growers using municipal water in Massachusetts, including water originating from the Quabbin Reservoir.

Of greater concern is the case where water having both high pH and high alkalinity is used for irrigation. In Massachusetts this situation is most common in Berkshire County. One reason is that the pH of the growing medium tends to increase significantly with time. In effect the water acts as a constant and dilute solution of limestone! This increase may be so large that normal lime rates must be reduced by as much as 50%. The problem is most serious when plants are grown in small containers because small volumes of soil are poorly buffered to pH change. Therefore, the combination of high pH and high alkalinity is of particular concern in plug and seedling trays. Trace element deficiencies such as of iron and manganese and imbalances of calcium (Ca) and magnesium (Mg) can also result from irrigating with high alkalinity water.

Potential Beneficial Effects on Nutrition

Water with moderate levels of alkalinity (30-60 ppm) can be an important source of Ca and Mg for some greenhouse operators. With the exception of a few fertilizers, many water soluble fertilizers do not supply Ca and Mg. Also, the Ca and Mg from limestone may be inadequate for some plants. Moderately alkaline water can beneficial as a source of extra Ca and Mg for crops prone to Ca and Mg deficiencies.

Effects of High pH and High Alkalinity on Pesticides

In addition to nutritional disorders of plants, water with high alkalinity can cause other problems. Bicarbonates and carbonates can clog the nozzles of pesticide sprayers and drip tube irrigation systems with detrimental effects. The activity of some pesticides, floral preservatives, and growth regulators is markedly reduced by high alkalinity. When some pesticides are mixed with water they must acidify the solution to be completely effective. Additional acidifier may be needed to neutralize all of the alkalinity.

If water pH is above 7.0, and the chemical requires a lower pH, a buffering (acidifying) agent should be added to lower the pH of the water used for mixing. Buffering agents can be obtained from greenhouse and nursery supply companies. Buffering agents should not be used with pesticides containing fixed copper or lime such as copper sulfate, or lime sulfur. Too much buffer should not be used as it may cause the water to become too acid and phytotoxicity may result. A pH of 6.0 is satisfactory for most pesticides.

To determine if a pesticide is affected by high pH or high alkalinity, carefully review the product's label. A call to the manufacturer may be needed to find the information for some chemicals.

Adjusting Alkalinity with Acids

Acidification of water having high pH but low alkalinity is rarely necessary, but many greenhouse operators inject acid (e.g., phosphoric, nitric, or sulfuric acid) into water with problematic high levels of alkalinity. The use of acid injection should be considered very carefully for several reasons. First, it is an extra step in production which will require additional materials and equipment. Second, acids are dangerous to handle and may damage some injectors and piping systems. Third, phosphoric or nitric acid are sources of P and NO_3, so the regular fertilizer program may need to be modified to take into account the addition of these nutrients. This would depend on how much acid must be used to neutralize the alkalinity and reduce pH. Finally, sometimes acid injection causes the solubilization of normally precipitated (unavailable) forms of trace elements, resulting in levels that are toxic to plants.

The amount of acid required to reach the desired pH (i.e., neutralize alkalinity) is determined by laboratory titration of a water sample with the appropriate acid or by a calculation procedure. Some "fine-tuning" may be needed later when actual injection is started. Acid is always added to water prior to the addition of fertilizer or other chemicals.

Acids have been and always will be an excellent tool for growers to exert better control of irrigation water alkalinity (mostly bicarbonates and carbonates) and growing media pH. Once the role of alkalinity is understood, the grower may consider the following practical steps to control alkalinity using acids through an injector system.

Acid Type

The tommonly available to growers include phosphoric, sulfuric, nitric, and citric. lists criteria for choosing the right acid for your situation: relative safety, neutralizing power, cost, and nutrient content. One of the most widely used acids is sulfuric acid; however, this is one of the most hazardous acids to use. For low amounts of alkalinity removal, phosphoric acid may be a better option. However, adding more than 2.25 fluid ounces of phosphoric acid to 100 gallons of water is not recommended, because of the amount of P that would be added. Nitric acid is theoretically ideal because it adds nitrate nitrogen; but it fumes and is highly oxidizing, making it very difficult and potentially dangerous to handle. Citric acid is a weak organic acid and a solid, making it safer than the other three; but it is much less effective, and therefore more expensive to use.

Using the Correct Injector

Once you choose an acid to use, make sure your injector can handle the task. Read the injector manual to get this information or call the manufacturer of the injector. Note: Some injector manufacturers state that a maximum of 5 percent acid can be used. This equates to approximately 6 fluid ounces of acid/gallon of water - an uncommonly high concentration of acid.

Calculating the Amount of Acid to use

It is suggested to use enough acid to reduce water alkalinity to within a target range. provides suggested target alkalinity ranges based on container size. First, have your water analyzed for alkalinity. You can have a lab test your alkalinity or you can use a kit to measure it yourself (alkalinity test kits can be purchased through greenhouse or scientific supply distributors). Then, calculate the amount of acid needed to get the water into your target alkalinity range. (Current alkalinity - desired alkalinity = alkalinity to be neutralized).lists the amount of acid to use for a certain ppm (parts per million) of alkalinity per 100 gallons of water.

Safety First

Acids are hazardous chemicals. When concentrated acids are mixed with water, a tremendous amount of heat is generated (which can even distort or melt plastic). Improper mixing can result in bodily injury. Always wear the proper safety equipment when using acids. This includes safety glasses, face shield, respirator, rubberized apron or coveralls, and acid-resistant gloves and boots. You should be able to find safety equipment distributors in the Yellow Pages under “Safety”. Federal and state safety laws and codes should be followed for storing, mixing and handling acids.

Proper Mixing

Use acid-resistant containers for containing the acidic stock solution. Heavy duty polyethylene trash cans are adequate. Always mix acid to water. Fill the stock container to about half the final volume you wish to mix with water. (Note: Since this is a pilot run, you do not want to make up a full amount of acidified stock solution because you may wish to adjust the amount of acid or to add fertilizer to the stock solution later.) Measure the acid carefully using a good measuring vessel. Then add acid to water, slowly and carefully to the center of the water surface. If dispensing acid from a large drum or container, you should invest in an acid-resistant, hand-activated pumping/dispensing device ("Industrial Suppliers in the Yellow Pages"). During and after adding acid to the water, you must stir the acid in the water. Acid is heavier than water, so don't assume it will mix easily just because it's a liquid.

Pilot Run or Calibration Measurement

After you have prepared the acidified stock solution, you should then determine if you have attained the target irrigation water alkalinity for your application. Run the injector at the appropriate dilution ratio for 5 to 10 minutes, then take a sample. It's best to run the water you wish to test into a 5-gallon bucket and take a sample out of the 5-gallon bucket. Test the alkalinity. Make adjustments as necessary. Once you are done, it is prudent to send another sample of the acidified water to an analytical lab to obtain a full test. This informs you if anything else has changed besides alkalinity.

Fertilizer Compatibility

Many growers want to use one injector and mix acid with fertilizers. The use of sulfuric, nitric and citric acid is compatible with most water-soluble fertilizers. Phosphoric acid is not compatible with calcium-containing fertilizers like calcium nitrate or formulations like 15-0-15 and 17-0-17 in concentrated form.

Mixing Fertilizer with Acid

If you are diluting the acid out of a separate injector, disregard this step. Remember, you only put in some of the acid to carry out the calibration run (half volume of stock solution). Add the remainder of the acid for the total amount of acidified water you wish to make. You may add more water, allowing "room" for fertilizer addition. Add the fertilizer carefully to avoid splashing, and add enough water to attain your final volume - mix thoroughly. Again, test the injection of the acidified nutrient solution to make sure the irrigation water is within the target alkalinity range.

Soluble Salts

Soluble salts in water are measured by electrical conductivity (ECw) expressed as millimhos per centimeter (mmhos/cm), which is equivalent to milliSiemens per centimeter (mS/cm). Electrical conductivity is also referred to as specific conductance or salinity.

EC (electrical conductivity) measures the levels of natural salinity and salinity caused by fertilizer residues in water and soils. In Massachusetts high EC water is not a common problem. However, high EC may occur in water from containment ponds rich in fertilizer residues, certain wastewaters used for irrigation, water contaminated by road salt, and rarely from saltwater intrusion in coastal wells. Irrigation water to which water-soluble fertilizer has been added has an EC of about 1.5-2.5 mS/cm, so, to avoid plant injury, the untreated water should have an EC no higher than the acceptable range of 0-1.5 mS/cm, although values of less than 1 are recommended for plugs. Excess soluble salts impair root function, which can lead to reduced water uptake and nutrient deficiencies.

Hardness – Calcium and Magnesium

Hardness is an indication of the amount of calcium and magnesium in the water. Calcium and magnesium are essential elements for plant growth that are reported in parts of element per million parts water (ppm) on a weight basis. Calcium in the range of 40 - 100 ppm, and magnesium in the range of 30 - 50 ppm are considered desirable for irrigation water.

Sodium and Chloride

Irrigation water from rivers, streams, private wells, and private ponds may contain excess sodium (Na) and chloride (Cl). A University of Massachusetts water study found that water from shallow private wells or private ponds was most likely to contain elevated Na and Cl due to road salt contamination. The contamination was most acute when these sources were located close to a road or parking lot. Municipal water generally had acceptably low Na and Cl levels probably because road salt applications are reduced in areas close to public wells and reservoirs. In wells and ponds Na and Cl levels were highest in the spring when runoff from snowmelt was highest or in the summer when water levels were drawn down to low levels during droughty periods. To properly assess how serious the Na and Cl contamination is, a series of water tests should be run during these periods to determine how high the levels are and their duration. This information will be useful in deciding what remedial measures to take.

While the most likely source of Na and Cl in the Northeast is road salt, water softeners and some fertilizers may also be contributors.

Na and Cl can be directly toxic to plants, may contribute to raising the soluble salts (EC) level of the growing medium, or may inhibit water uptake by plants. Plant problems include injury from excess soluble salts, growth reduction, and increased susceptibility to disease. Foliar chlorosis caused by high Na and Cl is similar in appearance to that caused by deficiencies of nitrogen, iron, and magnesium.

High sodium acts to inhibit plant uptake of calcium, and may result in excess leaching of calcium and magnesium from growing media. There is also the possibility of foliar absorption of sodium, resulting in leaf burn. Wells and municipal water sources may contain high chloride levels in association with sodium. The concern with chloride is the possibility of excessive foliar absorption under overhead irrigation or leaf edge burn caused by excessive root uptake in sensitive plants. If Cl concentrations are less than about 100 ppm, there is no concern from excessive foliar absorption. If Cl concentrations are less than about 150 ppm, there is no concern about toxicity resulting from root uptake. Increasing the level or frequency of water-soluble fertilizer should not be used as a corrective measure for these issues, as this only serves to further increase total EC and exacerbate the problem.

Acceptable levels of Na and Cl for ornamentals are less than 50 ppm and 140 ppm, respectively, however higher levels may be tolerated depending on crop sensitivity. Sodium levels of about 50 ppm or less are considered acceptable for overhead irrigation. Because of its effects on calcium and magnesium availability, the amount of sodium in irrigation water should be evaluated when you consider whether you have adequate calcium and magnesium. The effect of sodium is calculated as the sodium adsorption ratio (SAR). If the SAR is less than 2 and sodium is less than 40 ppm, then sodium should not limit calcium and magnesium availability.

If high levels of Na and Cl are suspected as plant problems, the suspicion should be confirmed by water testing every week or two weeks during the production season. If Na and Cl levels remain significantly above acceptable levels for weeks at a time then remedial measures should be considered. Changes in fertilization or other cultural practices do not provide long-term solutions for this problem. The best approach is to change salting practice to prevent contamination of the water source. Other remedies for road salt contamination are more expensive and involve changing water sources by drilling deeper wells away from roads and parking lots, switching to municipal water, or collecting and irrigating with rainwater. Another remedy would be water treatment by reverse osmosis. When feasible, municipal water, rainwater, or RO water can be mixed with saline water to dilute Na and Cl levels.

Potassium and Phosphate

These plant nutrients generally occur in water at very low levels. Presence in irrigation water at levels higher than a few parts per million may indicate the presence of pollution from fertilizers or other contaminants.

Sulfate

Sulfur is an essential element for plant growth that is not commonly included in fertilizers. It is measured in irrigation water to give an indication of possible deficiency problems. If the concentration is less than about 50 ppm, supplemental sulfate may need to be applied for good plant growth.

Ammonium and Nitrate

These nutrients are tested to give an indication of possible contamination of the water source. If present in significant amounts (e.g., >5 ppm nitrate), they should be taken into account in the fertility program. Fertilizer practices should be reviewed and corrected to prevent further contamination.

Micronutrients and Trace Minerals

The most important micronutrients are copper, zinc, manganese, iron and boron. They can occur in water sources in excessive or deficient quantities.

Excess iron and manganese compounds may result in unsightly residues on foliage under overhead irrigation. Fluoride may be present in levels high enough to damage foliage plants and Easter lilies. Concentrations in irrigation water should be less than 0.75 ppm. There may be a problem with the use of some fluoride-treated municipal water supplies.

Correcting Water Quality Problems

There are three major categories of water quality problems that can be corrected by chemical or physical treatment systems.

Alkalinity can be neutralized by addition of acids described in the alkalinity section above. Total dissolved solids, the soluble salts measured together as EC and individually in ppm of the element, can be removed by several water purification systems. Individual elements can be removed from the water if total dissolved solids are not high enough to warrant total salts removal. Before investing in any treatment system, however, it may be advisable to investigate the possibility of switching to an alternate water source, or mixing water sources, if it is an economical alternative for solving a water quality problem. Water purification methods and their applications are summarized in.

Water Purification to Remove Total Dissolved Solids

Two systems to remove total dissolved solids are reverse osmosis and deionization. Distillation and electrodialysis are water purification processes that can produce very high quality water, but at an often prohibitive cost.

Reverse Osmosis

This type of system removes 95 to 99 percent of the total dissolved salts. The system works by osmosis, which is the passage of a solvent (water) through a semi-permeable membrane separating two solutions of different salts concentrations.

A semi-permeable membrane is one through which the solvent can pass but the solutes (salts) can not. If pressure is applied on the solution with a high salt content (the irrigation source water), the solvent (water) is forced to move through the membrane leaving behind the salts. Relatively pure water accumulates on the other side of the membrane.

Maintenance and replacement of membranes are a significant part of the cost of reverse osmosis systems. Less efficient and less costly membranes are available that require less energy because of their lower operating pressures.

The amount of purified water delivered in a given time and the degree of salts removed depends on the pressure of the system, membrane type, total dissolved solids of the water being purified and temperature. Efficiency is strongly dependent on the integrity and cleanliness of the membranes. Chlorine can cause rapid degradation of the membranes and sediments cause clogging. For this reason, water to be purified by RO is usually pretreated to remove suspended solids, calcium carbonates and chlorine, and the pH is adjusted down if it is above 7.

Although total salts removal can be 95 to 99 percent, individual salts are removed with varying efficiency. In general, calcium, magnesium and sulfate are removed more efficiently than potassium, sodium, lithium, nitrate, chloride and borate.

A disadvantage of reverse osmosis systems is that salty wastewater is produced. Disposal of this waste may fall under government regulation.

Deionization

The soluble salts in water carry a charge that is either positive (cations) or negative (anions). Examples of cations are: sodium (Na^+), calcium (Ca^{++}), magnesium (Mg^{++}), iron (Fe^{++}) and potassium (K^+). Examples of anions are: chloride (Cl^-), sulfate (SO_4^-), bicarbonate (HCO_3^-), and fluoride (F^-).

Deionization is a process that removes ions from water using exchange resins. These are usually solid beads that are covered with fixed negative or positive charges. A cation exchange resin has fixed negative charges that are neutralized by H^+. When the irrigation water is passed over the resin, cations in the water replace the H^+ ions and are held on the resin. Likewise, an anion exchange resin has fixed positive charges that are neutralized by hydroxide ions (OH^-). When the irrigation water is passed over the resin, anions in the water replace the OH^- ions and are held on the resin. The H^+ and OH^- ions released from the resins combine to form water. A deionization unit will contain both anion and cation resins so that all salts are removed.

Deionization is very effective and produces a higher quality water than is generally needed in crop production. The cost increases with the amount of salts in the water to be removed. The higher the salts content, the more frequently the resins need to be regenerated or replaced. Cost of deionized water is generally five to six times higher than that of water purified by RO. If high-quality water is required (as for holding cut flowers) and the initial salts content of the water is high, RO can be used as an initial purification step and final quality be achieved by deionization. Final costs may actually be lower than with deionization alone.

Removing Individual Salts

Iron and Manganese

Iron and manganese in water become oxidized to insoluble forms that are responsible for black or brown stains on foliage of plants that are overhead irrigated. Iron concentrations of less than 0.3 ppm are required for micro-irrigation systems.

There are several ways to remove these elements. If enough space is available, the least expensive approach is to pump the source water into a pond or tank where the insoluble iron and manganese compounds can precipitate and settle out. The water is often pumped in as a spray for rapid oxidation of the iron and manganese to insoluble forms.

Enough time must be allowed for the iron and manganese to settle out, and the holding pond or tank must be large enough to accommodate the irrigation volume needs of the facility without disturbing the bottom layer of sediment containing the iron and manganese.

Oxidation filters also oxidize the iron and manganese to insoluble forms using air, potassium permanganate or chlorine. The sediments are removed by filters that must be periodically cleaned, usually by backflushing. Sand may also be used as a filter. If a chemical oxidant is used, this must be renewed as it is used up. Manganese is slower to oxidize and settle out of the water.

For efficient removal of manganese, chemical coagulation before sedimentation and filtering may be required. If iron and manganese bacteria are present, oxidizing filters should not be used. The oxidizing filters will be quickly blinded by the bacteria. Another approach to eliminating problems of precipitates is to keep the iron and manganese in soluble form. Polyphosphate chelates added to water attach to the soluble iron and manganese and keep them from becoming oxidized. The chelate-iron (-manganese) complex then passes through the irrigation system and is not precipitated on plants. Chelation generally works if the soluble iron and manganese concentration in the water is low (less than 1 - 2 mg/l). Furthermore, iron and manganese in the water that has already been oxidized by exposure to air cannot be chelated.

Water to which chelates have been added cannot be heated, because heating causes the polyphosphates to break down and release the iron and manganese.

Calcium and Magnesium

Calcium and magnesium may need to be removed from hard water to eliminate salt deposits left on foliage by overhead irrigation. This can be achieved by water softening; that is, replacing the calcium and magnesium with potassium.

Note that the usual water softening unit uses sodium, not potassium. High levels of sodium may be harmful to plants and a softening unit that uses potassium should be used instead. Total salt content of the water is not changed and the potassium is used by the plants. Over fertilization with potassium may occur if the water is very hard. The potassium chloride in the softening unit must be recharged.

Fluoride

Fluoride can be removed from irrigation water by adsorption using activated alumina or activated carbon. When using activated alumina, the pH of the water is first adjusted to 5.5 before treatment. The activated alumina unit can be regenerated with a strong base, such as sodium hydroxide, and reused. Water pH does not have to be adjusted before treatment with an activated carbon unit and the carbon is usually replaced when its adsorption capacity is used up. Fluoride is not soluble above pH 6 so maintaining a media solution pH above this level will prevent most fluoride toxicity problems.

Boron

Boron occurs in many irrigation water sources in the anionic borate form. Anion exchange resins similar to those described for deionization systems can be used, but at considerable expense. To increase the boron-removal efficiency of a reverse osmosis system, the pH of the water needs to be adjusted to be slightly alkaline (pH 7.5). Thin film composite type membranes that are more tolerant of the higher pH should be used.

Blending with Rainwater and other Non-problem Water

Rainwater can be collected from roof runoff structures, such as greenhouses, where it is then stored in a cistern to be used as irrigation water. Collected rainwater could also be blended with problem waters such as those with high alkalinity, high EC, or excess Na and Cl or to improve the quality of recycled tailwater and industrial wastewaters with high nutrient content used for irrigation. Other non-problem sources of water could also be used for blending. Rainwater has a natural pH of about 5.6 and a very low mineral content. Acidic rainwater with a pH in the range of 4.0-5.0 is acceptable for irrigation; it's poorly buffered and will have little effect on growing medium pH. Water with a pH below 4.0 should not be used as it may injure seedlings and young transplants. Rainwater should be collected from clean, well-maintained structures free from mineral contaminants such as zinc and other metals. Water should be tested for pH and minerals at least twice a year.

Water Management in Crop Production

Water is one of the most important inputs essential for the production of crops. Plants need it continuously during their life and in huge quantities. It profoundly influences photosynthesis, respiration, absorption, translocation and utilization of mineral nutrients, and cell division besides some other processes. Both its shortage and excess affect the growth and development of a plant directly and, consequently, its yield and quality. Rainfall plants. In India, however, rainfall is notoriously capricious, causing floods and droughts alternately. Its frequency distribution and amount are not in accordance with the needs of the crops. Artificial water supply through irrigation on one occasion, and the removal of excess water through drainage on another occasion, therefore, become imperative, if the crops are to be raised successfully. Water management in India, thus, comprises irrigation or drainage or both, depending considerably on the environmental conditions, soil, crops and climate. It is a situation oriented entity.

Water affects the performance of crops not only direclty but also indirectly by influencing the availability of other nutrients, the timing of cultural operations, etc. Water and other production inputs interact with one another. In proper combinations, the crop yields can be boosted manifold under irrigated agriculture.

Water is a costly input when canals supply it. The constructing of dams and reservoirs, the conveying of water from storage points to the fields, the operating and the maintaining of canal systems involve huge expense. The misuse of water leads to the problems of water logging, salt imbalance, etc. thus rendering agricultural lands unproductive. Hence a proper appreciation of the relationship and economic utilization of water resources for maximum crop production.

Water is one of the most important inputs essential crops. It profoundly influences photosynthesis, respiration, absorption, translocation and utilization of universal nutrients and cell division besides some other processes. Both it's shortage and excess affects the growth and development of the plants, yields and quality of produce. Rainfall is the cheapest of source of natural water supply. Its distribution is very uneven and uncertain. Artificial water supply through irrigation on one occasion and removal of excess water through drainage on the other becomes imperative. Water management in India thus comprises of irrigation and drainage or both. With proper combination of water and soil nutrients the crop yields can be boosted manifold under irrigated agriculture. Water is a costly input when canals supply it misuse of water leads to water logging, salt imbalance etc. rendering agricultural lands unproductive. Proper appreciation of the relationships among soils, crops, climate and water is essential for an efficient and economic use of water resources for maximum crop production.

Sources of Water

There are two sources of water

- Canal or surface water including lifts.
- Ground water-dug wells and bore wells.

Canal Water

This water becomes available from rivers after construction of dams on them and storing it in those dams. In India, rains are received during specific season from June to September called rainy season and in rest of the period there are no rains. This causes the problem of shortage or non-availability of water not only for irrigation but also for drinking. Therefore, artificial storaes are created by constructing dams across rivers. The rivers are small to big and accordingly dams of various sizes and capacity are built. They are mainly classified in three categories viz.1) minor 2) medium and 3) major or large:

- Minor-command less than 5000 ha.
- Medium-Command 5000 to 10,000ha.
- Major-Command more than 10,000 ha.

Although the command of minor dams is small, their benefits are accrued in short period. The designing, planning and construction of large dams takes long time and hence there benefit is available after a very long period. There are controversies over small Vs large dams. Water from these dams is made available for drinking and other human purposes, for industries and for irrigation. Huge capital operating and administrative costs are involved in making dam water available for above-mentioned purposes. Therefore, canal water is a costly input. Few dams were constructed during British Period before Independence. But looking to the pressing needs for water large scale programmers of development of river dams were undertaken on priority basis. By now thousands of dams have been constructed, thousands of Crores of rupees have been spent and several lakhs of hectares of land has been brought under irrigation, besides providing water for drinking in cities and towns and for industrial other uses. Actual are irrigated is 12 lakh ha. In Maharashtra alone and 640 lakh ha. In India.

Well Water

Well is traditional source of water. Wells are mostly constructed by individual farmers from their own funds or bank loans as it does not require much technical know-how. Wells can be completed in a short period and their benefits are available immediately. The farmers have full control on the use of waste for crops. They decide what crops to grow, when to irrigate and how much water to apply. There is generally no misuse or excess use of well water and hence no spoilage of lands. The command of each well is very small (5 to 10 ha) but the number of wells is very large. Therefore total area irrigated by wells is substantial. In Maharashtra nearly 65% of irrigated is under well irrigation. Due to uncertainty of rainfall, ground water recharging and its availability in wells is uncertain. But since farmers have control on use of water, delicate crops or crops requiring frequent watering but less water at a time, are grown on well water. Fruit, vegatables and differences in cropping patterns under canal and flower crops are commonly grown on well water. There are differences in cropping patterns under canal and well irrigations.

Water Requirements of Crops

Although water is essential for crop growth, it's requirement for each crop is definite. It may change according to the season (Kharif, Rabi or Summer) or according to soil type (light, medium and heavy) but net requirement may not change. The total water requirements, critical stages of water application and number of irrigation turns through which this water is to be given have been decided by experimentation. This information is given below:

Season and Crop	Water Requirement	No. of turns
I. Kharif		
i. Rice	50	6
ii. Jowar	20	3
iii. Bajri	20	3
iv. Maize	30	4
v. Pulses (Tur)	20	3
vi. Groundnut	40	5
vii. Sunflower/Soybean	25/25	3/3
viii. Chilli	40	5
ix. Cotton (Irri)	90	10
x. Onion	40	5
xi. Vegetables	40	5
II. Rabi		
i. Wheat	40	5
ii. Gram	30	4
iii. Groundnut	60	8
iv. Sunflower	40	5
v. Safflower	40	5
vi. Maize (Fodders)	60	8
vii. Vegetables	60	10
viii. Onion	60	10
III. Summer		
i. Groundnut	80	10
ii. Sunflower	60	6
iii. Maize (Fodder)	80	10
iv. Vegetables	80	10
v. Onion	70	9

IV. Perennial		
i. Sugarcane		
a. Adsale	400	32
b. Suru	300	25
ii. Fodder crops (Berseme/lucern,etc)	200	25
iii. Banana	240	24
iv. Other fruits	200	25

Irrigation Layouts

The water requirements of different crops are different, their root systems and feeding zones are different and the soils are also of different types. Therefore, following types of layouts are used for irrigating these crops. 1. Ridges & furrows - e.g. Sugarcane and different types of vegetables, chillies. 2. Long beds (Saras) - e.g. Wheat, Jowar, Bajra, Groundnut, Maize, etc. 3. Small beds - e.g. Onion, Potato. 4. Rings - e.g. Gourds (Vine crops) 5. Raised beds - e.g. Raising seedlings of various crops. 6. Trenches - e.g. grapes, banana.

Preventing Water Losses

Water losses from the fields can be prevented by various techniques such as:

1. Cultural methods.
2. Minimum weeds or no weeds.
3. Mulching - Straw mulch, black plastic film mulch, pebel mulch, etc.

References

- Irrigation, biology: byjus.com, Retrieved 5 July, 2019
- Irrigation, technology: britannica.com, Retrieved 12 April, 2019
- Surface-irrigation, irrigation, water-resource-engineering: civiltoday.com, Retrieved 14 February, 2019
- Types-of-surface-irrigation-with: cststudy.blogspot.com, Retrieved 19 May, 2019
- Sprinkler-irrigation-system, farm-machinary, agri-inputs, agriculture: vikaspedia.in, Retrieved 9 March, 2019
- How-exactly-does-drip-irrigation-work: interestingengineering.com, Retrieved 28 August, 2019
- Drainage: britannica.com, Retrieved 17 January, 2019
- Surface-drainage, managing-wet-soils, dairy, agriculture: vic.gov.au, Retrieved 2 May, 2019
- Subsurface-drainage, design-parameters: pavementinteractive.org, Retrieved 22 August, 2019
- Types-of-subsurface-drainage-systems, managing-wet-soils, dairy, agriculture: vic.gov.au, Retrieved 20 February, 2019
- Water, water-management: indiaagronet.com, Retrieved 7 March, 2019

Chapter 5

Cultivation Practices of Important Field Crops

The processes which are associated with the production of plant-based crops are known as cultivation practices. There are many types of field crops which are cultivated such as cereals, pulses and fibre crops. The chapter closely examines the common cultivation practices to provide an extensive understanding of the subject.

Cultivation Practices of Crops

1. Linseed

    - Propagation: Flax and linseed are mainly grown from seed and rarely vegetatively from stem cuttings.

2. Soil Preparation

    - Conventional Tillage: Because of its small seed size, linseed requires a fine, firm seedbed. Plant in moist soil and avoid sowing deeper than 5 cm. Avoid crusting soils because of low seedling vigour.
    - Minimum and Zero Tillage: Minimum and zero-tillage practices are attractive, because these provide protection from erosion, increase the soil organic matter and improve moisture retention. In addition, crusting problems and sun scald are reduced, consequently improving the stand of flax.

Planting

Flax is normally planted in the winter rainfall areas from mid-May to mid-June after the first winter rains. Planting time is very important and late planting (later than mid-June) can reduce the yield considerably.

Flax should be sown shallowly, 2,5 to 4,0 cm deep, in rows 15 to 20 cm apart. In the case of zero-tillage practices a row spacing of up to 30 cm is acceptable.

Fibre flax varieties are sown at 65 kg/ha, while linseed is sown at 50 kg/ha. This lower plant population allows the plant room to form an abundance of branches.

After reaching the two-leaf stage and hardening by exposure has taken place, seedlings can withstand temperatures as low as -4 °C for short periods without damage.

Fertilisation

Soil tests and experience should guide fertilisation practices. Nutrient levels in the soil vary greatly among regions and with soil types, cropping history and fertiliser use. Flax is very sensitive to fertiliser applied with the seed and even low rates sometimes cause seedling damage.

Nitrogen

Flax often responds well to the application of nitrogen (N) fertiliser when the available soil N is low. For flax, an N rate of 35 to 80 kg/ha should be used, unless more specific information, such as a valid soil test, indicates otherwise. Select a rate from within this range, based on moisture conditions. Use the lower end of the recommended range when soil moisture is low and the yield is expected to be limited. Use the higher recommended range when the soil moisture has to be maintained at optimal levels. An important rule is not to place the N directly with the seed.

Phosphorus

Apparently the flax plant prefers high soil P levels obtained from the fertilisation of preceding crops to high P levels obtained from the application of a high rate of P to the flax crop itself. Rates of about 35 kg/ha are recommended on soils with a low P content. The P must not be placed in direct contact with the seed.

Other Nutrients

- Fe – iron-deficient flax is chlorotic early in the season and usually appears as irregular patterns on the field. Such symptoms may disappear later in the season.
- Zn – zinc-deficient plants are chlorotic and the growing tips tend to die, whereafter axillary buds develop to form many branches.

Irrigation

Flax is an excellent choice for irrigated crop rotation because it is not prone to Sclerotina stem rot that affects canola, sunflower, peas and beans. The major effect of irrigation on flax is to promote a second or third flush of flowers and to maintain adequate moisture for plant growth until all the flowers have developed seeds.

The result of insufficient irrigation on flax

In nonrestricting soils (medium-textured soils that are amenable to plenty of moisture) flax develops a short, branched taproot, encompassing a rooting zone of 1 m. Root development is nearly completed by the flowering stage. In irrigated lands, flax takes approximately 70 % of its water requirements from the top half of the root zone.

Over the growing season, crop water use can be as high as 750 mm. During the seedling stage, water use will range from 1 to 3 mm per day, rising to a high of 7 mm per day during the flowering stage. The critical water requirement period is from flowering to just prior to seed ripening. Therefore, to maximise yield and oil content, adequate soil moisture must be maintained during this period.

Weed Control

Weeds can be a serious problem in flax if left uncontrolled. Early removal of weeds is necessary to minimise crop losses as a result of weed competition. Weeds are controlled more easily by herbicides in the seedling stage than at any other growth stage and early treatment usually decreases the risk of damage to the flax crop.

Weeds in flax field

Pest Control

Flax may be infested by various insect pests from the time of emergence to maturity. To keep damage low, fields should be examined regularly and control applied when infestations reach the economic threshold.

American bollworm

Earth and sand mites can cause damage to newly planted seed and to seedlings. Sowing pretreated seed can prevent this. Insect pests that may occur on locally cultivated flax are bollworms, ants and stainers. Bollworms can cause serious yield losses when feeding on the young fruit of the linseed plants.

American bollworm

Disease Control

Seed must be treated with a recommended fungicide. Seed treatment reduces seed decay and seedling blight, increases plant vigour and can increase the plant stand by up to 100 %.

Rust and Fusarium wilt are the two most frequent diseases associated with flax. Rust is recognised by the presence of a bright orange, powdery substance (pustules). Rust pustules develop on the underside of leaves and on stems and bolls.

Harvesting

Linseed

Linseed is considered to be fully matured when 75 % of the bolls have turned brown.

Linseed ready to be harvested

Flax may be harvested by straight combining or by cutting with a swather and threshing later with a combine or by hand. The latter is recommended as it assures drier seed for threshing.

Trials in the various potential linseed production areas in the Western Cape and Eastern Cape have shown that some varieties give yields of more than 2 tons of linseed per hectare.

Fibre Flax

Three stages are recognised in the ripening process of fibre flax:

- Stems totally green – when pulled at this stage the fibres are very fine but also very weak.
- Stems yellow – the fibres are long and supple at this stage and most suitable for processing.
- Stems brown – at this stage stems are strong but brittle; the fibres are too short and unsuitable for processing.

Fibre flax is harvested by a special pulling machine or may be pulled by hand. The flax is left in the field until dry, when the seed is threshed in such a way as to prevent breaking of the straw.

Fibre flax ready tobe harvested

Harvested fibre flax

Sunflower

- Propagation: Sunflower is propagated by seed.

Soil Preparation

Many different tillage systems can be used effectively for sunflower production. Soil preparation should be focused on decreasing runoff, especially in the case of soils with a low infiltration rate. These losses can be limited to a great extent by applying the correct soil cultivation practices. Conventional systems of seedbed preparation consist of mouldboard plowing or chisel plowing. The aim of the cultivation is to break up limiting layers, destroy weeds, provide a suitable seedbed and to break the soil surface and at the same time to ensure maximum rainfall infiltration, as well as to prevent wind and water erosion. Both germination percentage and lodging have been shown to increase in ridge-till systems vs. level plantings. Several tillage systems have been used with some success in specific environments. Major considerations are:

- Firm placement of seed near moist soil.
- Absence of green vegetation during emergence.
- Maintaining an option to cultivate.
- Reducing the risk of soil erosion.

Field Layout and Design

The field should be free of weeds, hardpans, stones and water logged conditions. Ridges, field waterways, terraces or contours and water harvesting basins should be constructed. Avoid fields that were planted to sunflower the previous year.

Planting

The planting density for sunflower ranges from 25 000 to 35 000 plants per hectare, depending on the yield potential of the area. Row width has little influence on grain yield. It can range from 90 to -100 cm, however, wider rows such as 1,5 m to 2,1 m can also be used, particularly to accommodate other farm implements. The required spacing in the row is about 30 cm.Sunflower seeds are planted at relatively shallow depths. In soil with a high clay content, seeds are planted at a depth of 25 mm. In sandy soils, seeds can be planted at a depth of up to 50 mm.

Fertilisation

Compared to grain crops, sunflower utilises soil nutrients exceptionally well. The main reason for this is the finely branched and extensive root system. The roots come into contact with nutrients which cannot be utilised by other crops.

Macro-nutrients

Sunflower normally reacts well to nitrogen and phosphorus fertilisation, where there is a shortage of these elements in the soil. It is therefore essential that any fertilisation programme for sunflower should be based on soil analyses. Soil analyses will not only lead to more appropriate fertilisation levels, but can also significantly limit unnecessary fertilisation costs.

Nitrogen

When there is a shortage, the growth rate decreases dramatically, leaves turn to pale green and the lower leaves die-off.

Table: Guidelines for nitrogen fertilisation (kg/ha).

Target yield (kg/ha)	N- guidelines (kg/ha)	
	Clay soils	All other soils
1 000	0	0–15
1 500	0–15	25–35
2 000	40–50	60–70

Phosphorus

A shortage of phosphorus is characterised by retarded growth. In serious cases necrosis can be detected on the tips of the lower leaves. Factors which should be taken into account when planning a phosphorus fertilisation programme, are the following:

i) Attempts should be made to build up the phosphorus content of the soil over time.

ii) The optimum soil phosphorus level for sunflower is about 10 mg/kg.

This means that phosphorus fertilisation is essential when the level of phosphorus in the soil is below 10 mg/kg. However, at a higher level, the crop will probably not respond to phosphorus fertilisation.

Potassium

Although sunflower draws large quantities of potassium from the soil, potassium fertilisation is usually unnecessary, as South African soils generally have adequate quantities of this nutrient.

Molybdenum and Boron

Shortages of boron and molybdenum often limit the growth and yield of sunflower in the eastern parts of the country. To avoid problems concerning these two elements, care should be taken to apply fertiliser containing boron and to ensure that seeds are treated with molybdenum prior to planting. Local seed companies usually treat their seed with molybdenum. If no soil analysis is available, 50 to 100 kg/ha of a 3:2:1 (25) fertiliser mixture applied at planting is adequate for a yield potential of 1 000 to 1 500 kg/ha.

Irrigation

In most cases South African sunflower is cultivated under dry land during the summer rainfall season (November to March). In areas with low rainfall, water supply can be supplemented with irrigation in order to increase yield. The method of irrigation will depend on the water availability and the available irrigation equipments. The pH of the irrigation water should be slightly neutral.

Weed Control

Efficient weed control is a prerequisite for high sunflower yields. It is achieved by a combination of mechanical and chemical practices. Young plants are highly sensitive to strong weed competition and cannot develop fast enough to form a full shade covering which can suppress weed seedlings. Therefore, the first 6 weeks after planting are a critical period for the crop.

Yields can be increased significantly by keeping fields free of weeds during this time.

Mechanical Weed Control

Mechanical weed control can be very effective, provided it is done in time and with care not to damage the crop. Chemical weed control can be applied successfully together with mechanical methods and cultivation practices to bring about better weed control.

The following tips are given for mechanical control:

- Cultivate before the sunflower is too tall for equipment, or the plants will be damaged easily.
- To prevent damaging the sunflower roots, cultivation should be shallow (less than 75 mm).
- Throw loose soil onto the row—this will assist in suffocating weeds which sprout in the row.
- Smaller weeds die-off easily when dry soil is hoed.
- Hoe during the hottest part of day when the sunflowers are wilted—this reduces stem damage.

Chemical Weed Control

The use of herbicides has many advantages, of which the most important is that effective weed control can be applied during wet periods when mechanical weed control is impossible. If sunflower is cultivated in crop rotation with maize, weeds can be controlled more effectively in both crops as grass and broadleaf herbicides can be used in continuous succession.

Pest Control

Insect pests have become major potential yield-reducing factors in sunflower producing areas.

Frequent Insects in Sunflower

These potential risks require that growers follow integrated pest management (IPM) practices. IPM means using a combination of pest control methods to maintain pest populations below levels which result in unacceptable losses to crop quality or yield. IPM includes biological, cultural, physical and chemical control. Resistance to insects can be improved by the presence of a dark- coloured "armour" layer on the seed coats. Resistance to midges has been suggested, however, it is not presently effective. Only currently approved insecticides should be used for control of insects. Birds can be major pests in sunflowers, especially blackbirds, goldfinches, doves, grosbeaks and sparrows. Many approaches to feed disruption have been tried, including scarecrows, fright owls, aluminum strips that flutter in the wind and carbide exploders. No technique is 100 % effective, as birds will adapt to many of these techniques. Currently, no chemicals are approved for bird control in sunflower.

Table: Control measures for pests on sunflower.

Pests	Pest description	Symptoms	Control measures
American bollworm	First instar larvaealmost black Colour of the older larvae can vary from nearly black to brownish, green or even pink Underside dirty white with a stripe along each side with very distinct spiracles.	Feeds mainly on young heads.	Crop rotation, grow resistant cultivars, use of certified seeds.
Astylus beetle (larvae)	Hairy grub of medium size.	Feeds mainly on decomposing organic matter,however, will also attack young seedlings.	
Black maize beetle	Shiny, black beetle approx-imately 12 mm long. Very active and a good flier.	Eats holes in plants below soil level just above germinating seed Larvae feed on dead organic matter in the soil.	
Nematodes	Microscopic slender transpar-ent worms in the soil.	Attacks roots of plants causing lesions that make these susceptible to bacterial or fungal attack, resulting in poor growth of plants.	
Weevils	Small, wingless beetles with distinct snouts Larvae are whitish and legless.		

Disease Control

The most serious diseases of sunflower are caused by fungi. The major diseases include rust, downy mildew, Verticillium wilt, Sclerotinia stalk and head rot, Phoma black stem and leaf spot. Resistance to rust, downy mildew, and Verticillium wilt has been incorporated into improved sunflower germplasm. The severity of these diseases effects on total crop yield could be ranked:

1. Sclerotinia
2. Verticillium
3. rust (recently more severe)
4. Phoma
5. downy mildew.

Disease Control Mechanisms

Disease	Symptoms	Control mechanisms
Downy mildew (Plasmopara halstedi)	Stunted growth, yellowgreen and green lesions on the leaves Cottony fungus on the underside of the leaves Blackening and swelling on the stem Severe in humid areas.	Crop rotation Conservation Tillage, scheduled irrigation Disease resistant cultivars Drought resistance cultivars Registered seeds.
Powdery mildew (Erysiphe cichoracearum)	Cottony fungus on green leaves late in summer (no large-scale damage).	
Leafspot (Septoria helianthi)	Dead lesions on flower leaves before heading Not causing appreciable loss.	
Verticillium wilt (Sclerotinia sclerotiorum)	Dead areas along leaf veins, bordered by light yellowgreen margins Decayed vascular tissue in cross-section of the stem.	
Rust (Puccini helianthi)	Rust coloured postules on leaves, latter black speck on stems.	
Sclerotinia head and stem rot (Verticillium dahliae)	Wilt soon after flowering Light tan band around the stem at the soil leve Grey-black sclerotia (seed size) in rotted heads and stems.	
Phoma black stem (Phoma macdonaldii)	Large chocolate coloured lesions on stems at maturity.	

Crop Rotation

Sunflower should be grown in rotation with other crops as:

- The risk of diseases and weeds increase with monocropping.
- A yield and quality advantage is often measured in a follow-up maize or sorghum crop.
- Weed and pest problems decrease with crop rotation.

However, take note that some herbicides do have a long residual period and could damage the follow-up crop in a rotation system. It is therefore important to strictly follow instructions on herbicide labels.

Harvesting

Harvest Maturity

Harvesting should commence as soon as 80 % of the sunflower heads are brown in order to minimise losses caused by birds, lodging and shattering. The leaves turn yellowish during harvesting maturity. Sunflower is generally mature long before it is dry enough for combining. The sunflower plant is physiologically mature when the back of the head has turned from green to yellow and the bracts are turning brown, about 30 to 45 days after bloom, and seed moisture is about 35 %. The total growing period (from seeding to harvesting) for sunflower ranges from 125 to 130 days.

Harvesting Methods

Harvesting is done either manually or mechanically. Manual harvesting is practiced by cutting the crop with a sickle or knife. Commercially available sunflower headers are useful in decreasing loss of seed as the crop is direct combined. The combines are used to perform several operations such as cutting the crop, separating the grain from the straw, cleaning the grain from chaff and transporting grains to the storage tank.

This equipment usually includes 20 to 90 cm width metal pans for catching matured seed and a three-armed or similar reel. A narrower (23 cm) pan width enables harvesting diagonal to the row, which produces smaller harvest losses in some situations. Harvesting with a combine harvester will save more time than manual harvesting.

Cultivation Practices of Cereals

Rice

Seed Quality and Selection

Seed is a living product that must be grown, harvested, and processed correctly in order to realize the yield potential of any rice variety. Good quality seed can increase yields by 5-20%. Using good seed leads to lower seeding rates, higher crop emergence, reduced replanting, more uniform plant stands, and more vigorous early crop growth. Vigorous growth in early stages reduces weed problems and increases crop resistance to insect pests and diseases. All of these factors contribute to higher yields and more productive rice farms.

Good seed is pure (of the chosen variety), full and uniform in size, viable (more than 80% germination with good seedling vigor), and free of weed seeds, seed-borne diseases, pathogens, insects, or other matter.

Choosing seed of a suitable variety of rice that suits the environment it will be grown in and ensuring the seed choosen of that variety is of the highest possible quality is the essential first step in rice production.

Land Preparation

Before rice can be planted, the soil should be in the best physical condition for crop growth and the soil surface is level. Land preparation involves plowing and harrowing to 'till' or dig-up, mix and level the soil.

Tillage allows the seeds to be planted at the right depth, and also helps with weed control. Farmers can till the land themselves using hoes and other equipment or they can be assisted by draft animals, such as buffalo, or tractors and other machinery.

Next, the land is leveled to reduce the amount of water wasted by uneven pockets of too-deep water or exposed soil. Effective land leveling allows the seedlings to become established more easily, reduces the amount of effort required to manage the crop, and increases both grain quality and yields.

Crop Establishment

The two main practices of establishing rice plants are transplanting and direct seeding. Transplanting is the most popular plant establishment technique across Asia. Pre- germinated seedlings are transferred from a seedbed to the wet field. It requires less seed and is an effective method to control weeds, but requires more labor. Seedlings may be transplanted by either machine or hand.

Direct seeding involves broadcasting dry seed or pre-germinated seeds and seedlings by hand or planting them by machine. In rainfed and deepwater ecosystems, dry seed is manually broadcast onto the soil surface and then incorporated either by ploughing or by harrowing while the soil is still dry. In irrigated areas, seed is normally pre- germinated prior to broadcasting.

Water use and Management

Cultivated rice is extremely sensitive to water shortages. To ensure sufficient water, most rice farmers aim to maintain flooded conditions in their field. This is especially true for lowland rice. Good water management in lowland rice focuses on practices that conserve water while ensuring sufficient water for the crop.

In rainfed environments when optimal amounts of water may not be available for rice production, a suite of options are available to help farmers cope with different degrees and forms of water scarcity. It includes sound land preparation and pre-planting activities followed by techniques such as saturated soil culture, alternate wetting and drying, raised beds, mulching, and use of aerobic rice that can cope with dryer conditions.

Nutrient Management

At each growth stage, the rice plant has specific nutrient needs. This makes nutrient management a critical aspect of rice farming.

The unique properties of flooded soils make rice different from any other crop. Because of prolonged flooding in rice fields, farmers are able to conserve soil organic matter and also receive free input of nitrogen from biological sources, which means they need little or no nitrogen fertilizer to retain yields. However, farmers can tailor nutrient management to the specific conditions of their field to increase yields.

Crop Health

The rice plant has a wide array of 'enemies' in the field. These include rodents, harmful insects, viruses, diseases, and weeds. Farmers manage weeds through water management and land preparation, by hand weeding, and in some cases herbcide application. Understanding the interactions among pests, natural enemies, host plants, other organisms, and the environment allows farmers to determine what if any pest management may be necessary.

Avoiding conditions that allow pests to adapt and thrive in a particular ecosystem helps to identify weak links in the pests' life cycle and therefore what factors can be manipulated to manage them. Retaining natural ecosystems such that predators and natural enemies of pests and diseases are kept in abundance can also help keep pest numbers down.

Harvest

Harvesting is the process of collecting the mature rice crop from the field. Depending on the variety, a rice crop usually reaches maturity at around 105–150 days after crop establishment. Harvesting activities include cutting, stacking, handling, threshing, cleaning, and hauling. Good harvesting methods help maximize grain yield and minimize grain damage and deterioration.

Harvesting can be done manually or mechanically:

i) Manual harvesting is common across Asia It involves cutting the rice crop with simple hand tools like sickles and knives. Manual harvesting is very effective when a crop has lodged or fallen over, however it is labor intensive. Manual harvesting requires 40 to 80 hours per hectare and it takes additional labor to manually collect and haul the harvested crop.

ii) Mechanical harvesting using reapers or combine harvesters is the other option, but not so common due to the availability and cost of machinery. Following cutting the rice must be threshed to separate the grain from the stalk and cleaned. These processes can also be done by hand or machine.

Maize

Choice of Suitable Cultivar

Seed is the driver of all investments and management practices by the farmer on a crop, and therefore, choice of right cultivar is of paramount importance for cultivation of a successful and profitable maize. Cultivars suited to particular agro-ecological region, season, purpose and maturity should be selected. It is better to select a single cross hybrid as its performance is superior under optimal conditions, and also under mild stress conditions. In case of known stressful or poor growing conditions, like low moisture availability, low input conditions etc., composites, open pollinated varieties (OPVs) could be used. Usually, there are plenty of hybrids available through various sources, including retail seed market, government seed outlets etc. In general, all these options are available with claim of high yields, and therefore it is sometime difficult for the farmers to make a right choice. Therefore, some basic criteria need to be used for selecting a suitable hybrid.

Suitability of a cultivar for a particular location (field) depends upon of water availability, availability fertilizer and other inputs, cropping season and cropping system window, as follows:

i) Maturity group: Choice of suitable maturity cultivar (early, medium or late/full-season) should be based on crop season (Rabi, Kharif or Spring) at a particular location depending on moisture availability regime. Cropping season window (period from planting to harvesting) need to be critically assessed before making a choice on the basis of:

- Cropping window available in the system - based on harvesting time of previous crop and planting time of next crop.
- Length of cropping season with no prolonged weather extremes, such as too high (>35 °C) or too low temperature (<5 °C), and
- Moisture availability – rainy days or irrigation facility in case of no-rains (and no excessive moisture/waterlogging at any crop stage).

On consideration of above factors, the available cropping window for choice of maturity group of hybrids, as follows in the table.

Approximate duration of cropping window	Suitable maturity group of cultivars
>100 days	Late maturity (Full-season)
90-100 days	Medium maturity
80- 90 days	Early maturity
<80 days	Extra-early maturity

ii) Moisture availability: It is one of the most critical criteria for selection of suitable cultivars. If water availability is adequate for whole cropping season based on good rains or supplemental irrigation facility (in case of no rains), then high yielding singlecross hybrids within selected maturity group is a good choice. However, in case the crop is to be grown under rain-fed conditions with erratic/uneven moisture conditions (without supplemental irrigation) then it is important to choose a stressresilient cultivars, depending on expected moisture regime; for example - a drought tolerant hybrids to cope-up with intermittent drought stress.

iii) Temperature regimes: A suitable temperature for maize crop for good growth, development and yield is 25– 35 °C. Temperature below or above this range (critical limit) may cause negative effects on maize crop. Especially under prolonged (>5 °C days) extreme temperature regimes, low (35 °C) during active growth stages (vegetative, flowering and early grain fillings stages), crop may suffer badly and eventually end-up with poor yields. In case of locations or cropping season with known extreme temperature regimes, cultivar with tolerance to such temperature extremes need to be selected. For example – in spring season (planting in February/March month) temperature is invariably high (>35 °C) in most part of the cropping season, therefore, for such heat stress prone environments, hybrid with in-built heat tolerance is must. Similarly, for the areas with prolonged low temperature period during winter season a hybrid with cold tolerance traits should be selected for cultivation.

It is advised to use certified seed of the recommended cultivar for specific agro-ecological situations. Seed of hybrids must be purchased fresh every year because use of saved grain (F2 seed) from the previous year crop likely to give low yield. However, in case of composites/open pollinated varieties (OPVs), farmers may save some selected cobs at the time of harvest for using as seed next year. In such case, uniform cobs with good seed setting, free from any disease or insect-damage, should be selected from centre of the field (avoid cobs from border sides). After shelling, seed should be properly cleaned, dried and stored in dry place using low humidity storage condition, such as polythene bags.

Planting Time

Maize can be grown round the year, viz. Kharif, Rabi and Zaid/Spring season. Timely planting is one of the key factors deciding the overall performance of the crop, and eventually grain yield. However, it varies with location-to-location, and therefore local recommendations for specific crop season for maize planting time need to be followed.

- Kharif maize: In general, in Kharif season maize planting should be completed with first monsoon rain of the season. However, if irrigation facility is available then sowing should be taken 1-2 weeks prior to on-set of monsoon. This will help crop to get longer duration of cropping season, which eventually helps in good yields, and also crop is harvested before the next season crop planting time. However, in rainfed area, the sowing time should coincide with onset of monsoon.
- Rabi maize: In winter season, planting time need to adjusted in a way that reproductive stage.In such areas, the suitable planting time is between 15 October – 15 November. Last week of October for intercropping and first fortnight of November for sole crop is the ideal time for rabi maize planting. In general, Rabi season planting in most of the areas should be completed within November month, as further delay will result in delay in flowering time, which may coincide with high temperature regimes in the month of March.
- Spring maize: During this season maize crop is invariably exposed to high temperature regimes at later crop growth stages in the months of April-May. Therefore, early planting is very critical to avoid critical crop stages, such as flowering and early grain-filling stages exposed to extreme heat condition(35° C or more). It is recommended that spring season maize planting should be completed latest by mid-February to avoid heat stress at critical reproductive stage. It also helps harvesting of the crop before on-set of monsoon rain. Late planting may cause various types of damage to standing maize crop including severe yield losses due to exposure to heat stress apart from grains with high moisture at harvest, and also delay in sowing of next crop in Kharif season.

Field Preparation and Planting

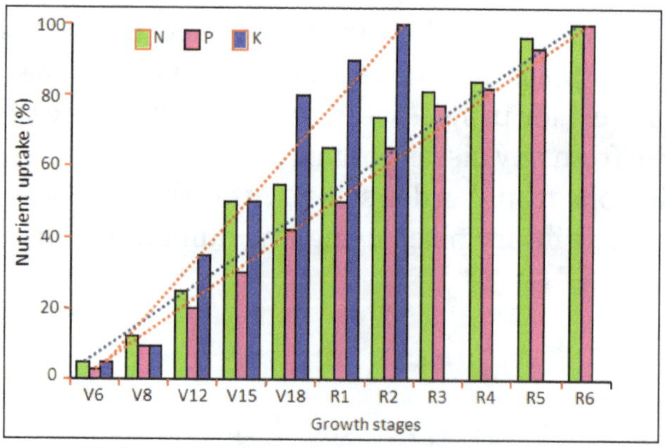

Maize under ridge-furrow planting system.

Well-ploughed field with pulverized soil is helpful in making ridges (narrow raised-beds) in the field for planting maize crop. Ridge-furrow system is ideal for planting maize crop to avoid excessive moisture/waterlogging patches after each irrigation (or rains) by providing adequate moisture : air ratio in the root zone. The other choice would to be plant on flat surface and earthing-up the crop immediately after 2nd top-dressing of nitrogen at about four weeks after planting. Seeds should be sown about 5.0 cm deep to ensure good seedling growth and vigour. It should be ensured that seeds have about 95% germination. Seed lots with less than 80% germination are likely to give poor plant stand. In case of poor germination, gap-filling can be done as soon as possible to ensure the uniformity within same field.

Seed Rate and Planting Geometry

Yield of maize crop is directly related to number of total plants per unit area (per acre or hectare). Optimum plant density is the key for achieving higher yields and resource use efficiency. High planting should be used only in case of high input management, as over-density with low inputs may negatively affect the crop performance. Seed rate depends on purpose, seed size, plant type, season, sowing method etc. The recommended seed rate to achieve the required plant population of ~70 thousand/ha for grain maize crop is 20.0 kg per hectare (or 8.0 kg/acre). To achieve this, it is necessary to attain about 10% higher stand at germination. Line sowing with suitable row-to-row and plant-to-plant distance is recommended for optimal performance of maize crop. Various planting geometries can be used, depending on purpose, soil-type, water availability and input management conditions for achieving higher yields.

Table: Seed rate and planting geometry of maize for different purposes.

Purpose	Seed rate (kg/ha)	Plant geometry (row x plant, cm)	Plant population
Grain	20	60 x 20 75 x 20	83,333 66,666
Green cob	20	75 x 20 60 x 20	66,666 83,333
Fodder	50	30 x 10	333,333

Seed Treatment

It is always advisable to treat the seed with fungicides and insecticides before sowing to protect the crop from major soil borne diseases and insect pests. The recommended seed treatment options are given in the table.

Table: Seed treatment for major diseases and pests.

Disease/insect-pest	Fungicide/Pesticide	Rate (g/kg seed)
Foliar diseases (Turcicum Leaf Blight,, Banded Leaf and Sheath Blight, Maydis Leaf Blight)	Bavistin + Captan in 1:1 ratio	2.0
BSMD	Apran 35 SD	4.0

Pythium Stalk Rot	Captan	2.5
Termite and shoot fly	Imidachlorpid	4.0

Nutrient Management

Proper fertility management is a key for growing a good maize crop and realizing the yield potential of a cultivar. Judicious and balance application of fertilizers is necessary to optimize returns on investment. Apart from chemical fertilizer, it is recommended that farmyard manure (FYM) @10-15 tonnes/ha should be incorporated into the field. Combination of organic manure and chemical fertilizer gives better yields as well as improve the soil fertility. Recycling of crop residue (dry maize stover) after harvest is also advised for improving soil health and for sustained nutrient supply.

i) Fertilizer doses: The precise quantity of fertilizers needed for various fields depend upon the fertility status of the soil, previous cropping history and the maturity group of the hybrid. It is advised that fertilizer dose should be decided on the basis of soil fertility information on soil testing basis (at least once in five year). Generally, fertilizer recommendation is agro-ecoregion specific and therefore recommendation by state agriculture university or agricultural department should be followed. A balanced application of all nutrients (rather than overdose of one and low-dose of others) is very important for proper crop growth and development, and good yields. Depending upon soil fertility status, maturity group of hybrids and soil type, 150-180 kg nitrogen, 70-80 kg phosphorous, 60-80 kg potash and 10 kg zinc per ha is recommended for maize crop. Early maturing varieties require lower quantity of nitrogen and fullseason hybrids may need higher dose.

ii) Time of application: In general, maize crop is very responsive to nutrients. The rate of nutrient application depends mainly on soil nutrient status/balance and cropping system. For achieving good yields, the nutrient application should be done in such a way that matches with the soil supplying capacity and plant demand (Sitespecific nutrient management approach) by keeping in view of the preceding crop (cropping system). Maize crop has different pattern of nutrient uptake for different nutrient during the crop cycle. For example – total K requirement is taken up by the crop by R_2 growth stage (about 2 weeks after anthesis), whereas uptake of N and P continued until R6 stage (near Physiological maturity) of the crop.Therefore, fertilizer application should match the uptake pattern of the nutrient by the crop. Though P and K are quite stable nutrient, may stay longer in the soil after application, however, N is quite unstable as either it will be either taken by the crop after application or it will be lost through various processes, like leaching, volatilization etc.

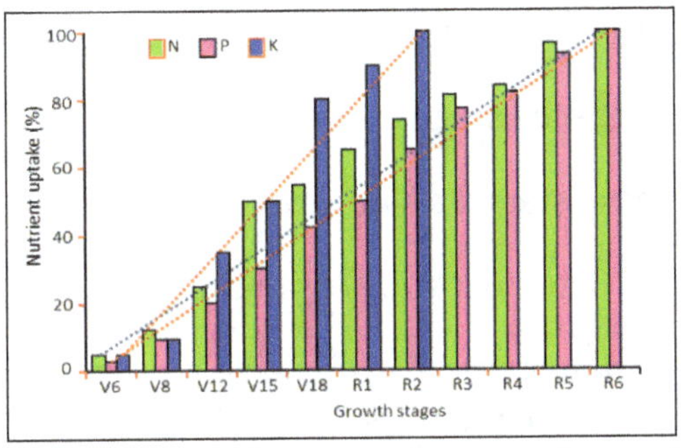

N, P and K uptake by maize plants at various growth stage in crop cycle.

Nutrient application, need to be planned in a way that the applied nutrient is matching with nutrient uptake pattern of the crop. Especially in case of N, which is quite unstable in soil, and crop N-uptake continued until R6 stage, the time of application and splits of total recommended dose of N need to be carefully matched with crop requirement, rather than over-application of N at early stages (which is mostly wasted), applying final dose at VT (tassel emergence stage) and later-on no N-application even if crop still needs some amount of N for successful completion of crop cycle.

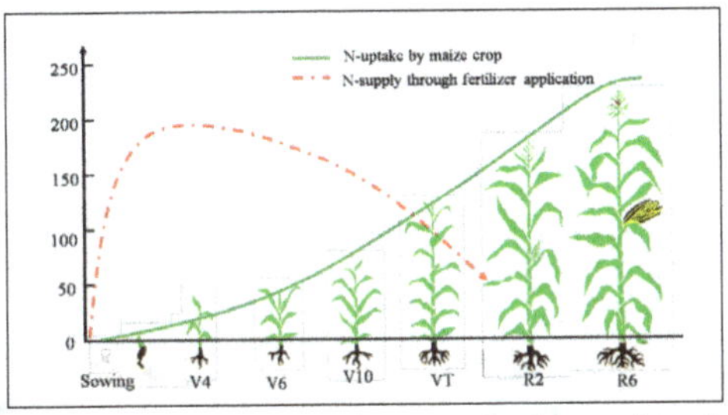

N-supply as fertilizer and N-uptake by maize plant.

Therefore, keeping view the nutrient uptake and its requirement pattern by maize crop, following scheduled of the recommended doses should be followed for enhancing nutrient use efficiency and better crop performance.

a) P-application : 100% as basal dose at the time of planting.

b) K-application: In three splits, including:

- 40% as basal dose at the time of planting.
- 40% at 5-6 leaf stage along with N application.
- 20% at tassel emergence stage, along with N application.

c) N-application: In at least five splits, including:

- 10% - as basal dose at the time of planting.
- 20% - at the time of 2-3 leaf stage as side dressing followed by irrigation (if no moisture in field).

- 40% - at the time of 5-6 leaf stage along with inter-culture and earthing-up.
- 20% - at tassel emergence stage as side-dressing.
- 10% - at silk browning stage as side-dressing.

d) Zn-application (if recommended): 100% as basal dose at the time of planting.

iii) Site-specific nutrient management (SSNM): Nutrient Expert is an easy-to-use, interactive computer based decision support tool that provides nutrient recommendation for individual farmers' and specific fields with or without soil test data. It synthesizes the on-farm research data into a simple delivery system that enables farmers to rapidly implement SSNM for their individual fields. The software estimates nutrient requirement based on attainable yield by combining information like fertilizer/ manure applied, crop residue retained in the previous crop, crop growing conditions etc. with expected N, P and K responses in target fields to generate location-specific nutrient recommendations for maize. The software also does a simple profit analysis comparing costs and benefits between farmers' current practice and recommended alternative practices. The algorithm for calculating fertilizer requirements was developed from on-farm research data and validated over five years of testing.

Moisture Management

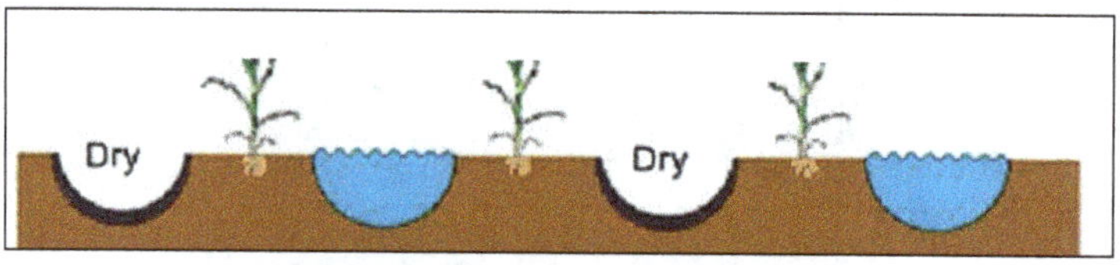

Half-irrigation by skipping alternate row

Furrow irrigation in maize field

Irrigation in maize field (a) furrow system, (b) half irrigation

Adequate moisture management is essential at critical growth stages for good crop growth and development. In general, irrigation should be applied in furrows, up to 2/3rd height of the ridges.Early seedling, kneehigh stage (6-8 leaf stage), reproductive stage (pollen sheding and silk emergence) and early grain filling (two weeks after flowering) are the most sensitive stages for water stress and hence good moisture should be ensured at these stages. In limited irrigation water availability conditions, the irrigation water can be applied in alternate furrows (skipping one furrow in between) in ridge-furrow planting system.Maize can be successfully grown under rain-fed conditions where the distribution of rainfall is enough to ensure adequate soil moisture during the crop cycle, and at least at critical crop stages mentioned above. In case of erratic rainfall distribution pattern, which is very common, one or more need-based supplemental irrigation may be required at critical growth stages, such as during 6-8 leaf stage and at the time of tasselling/silking stages. Though, severe drought conditions at any growth stage is undesirable for maize crop, reproductive stage especially during two weeks before to two weeks after anthesis is most critical, and therefore good moisture management at this stage is highly essential.

Being a tropical origin crop, excessive moisture/waterlogging negatively affects maize crop growth and development, as more than 24 hrs of water inundation may cause an irreversible damage to the crop. Therefore, proper drainage is necessary for a good maize crop. Sowings on raised bed (ridges-furrow system) with provision of surface drainage can overcome the constraints of temporary waterlogging after rains or irrigation.

Weed Management

(a) Cyprus rotundus & (b) Cynodon dactylon

During crop-free period, summer ploughing helps in reducing the existing weeds in field and also reduced the weed seed bank in the soil. In Kharif season, weeds are the serious problem in maize as they compete for light, nutrient, and water and causes yield losses up to 30-50%. Therefore, timely weed management is important for achieving higher yield. Atrazine being a selective and broadspectrum herbicide in maize checks the emergence of wide spectrum of weeds. Preemergence application (spray within 0-3 days after planting) of atrazine @1.0-1.5 kg a.i. ha-1 can successfully managed a broad-spectrum (both narrow and broadleaves) weeds except Cyprus and Cynodon. It will keep crop free from weeds at least one month until soil is disturbed at the time of 2nd top-dressing of N followed by earthing-up. At later stages, once the crop canopy is properly developed, the crop itself can suppress the newly emerging weeds. However, in case there is problem of weeds in standing crop at later stage, there are selective herbicides such as : Tembotrione (Trade name: Laudis @90g ha-1), Topramezone (Trade name: Tynzer or Armezon @80 g ha-1) or Halosulfuron (Trade name: Sempra @87.5 g ha-1) works well when applied along with atrazine (@1.0kg/ha a.i.).

One to two hoeing are recommended for aeration and up-rooting of the remaining weeds, if any. Now-a-days small machine such as 'power weeder' is available, which can be used for weeding and earthing-up in 20-25 days old maize crop which effectively controls diverse weed flora throughout the crop growth period in rain-fed maize. The machine has blades for tillage which cuts and mix weeds in soils and ridger or shaper attached behind the machine helps in earthing-up.

Weeding, inter-culture and earthing-up using power-weeder

Harvesting

(a) Maize kernels with black-layer at the base and (b) maize cobs with dry husk-cover, ready for harvest

Maize crop cultivated for grain purpose should be harvest any time after one week of physiological maturity. The physiological maturity could be identified by looking for a black-layer at the base of kernel in middle position of the cob. The black-layer is abscission layer, which indicates that the kernel are fully mature. Another easier way to identify the physiological maturity is when husk-cover (leaves wrapping around the cobs) is dried. At this stage kernel are fully mature, but moisture content in kernel might be 25-30%, and therefore, it is better to wait for another 2-3 week so that kernel moisture content come down to around 15%. Harvesting at low moisture content and storing in dry condition increases the longevity of grains in storage.

Depending on the purpose of the maize cultivation, maize crop can be harvested at other stages as well, for example - at late dough stage (about 5-6 week of after pollination) for green-cob purpose, and green fodder can be used for cattle feed.

Pearl Millet

Pearl millet is propagated by seed.

Soil Preparation

Prepare warmer seedbeds on well-drained soils. Seedbeds should be weed free. Deep till or in-row subsoil sandy textured soils are used to disrupt any hard pans. No-till or conservation-tillage plantings can be successful and are desirable on highly erodible land or clayey soils. This will reduce soil erosion and enhance stand establishment owing to better seed depth control in firmer soils and control of emerged weeds prior to planting. If no-tilling in the spring, deep tillage ahead of the winter cover crop in the fall is preferred. However, reconstitution of the hard pan in sandy soils can occur, particularly if good rainfall occurs during the winter.

Planting

Optimum planting time should be early October to November, and this is greatly dependent on the intended use. Soil temperatures should be at least 18 °C. Planting in cooler soils can cause problems with reduced emergence and greater competition from weeds. Plant densities should be similar or slightly higher (100 000 to 175 000 plants/ha) than for sorghum. Seed should be planted into a firm, mellow, moist seedbed. As pearl millet seed is small, shallow planting is recommended to obtain good seed-to-soil contact.

Fertilisation

Millets are generally grown on less fertile soils, but respond well to heavy fertilisation. They are deep-rooted and can use residual nitrogen, phosphorus and potassium and, therefore, may not need the levels of fertilisation required by other summer grains.

Because nitrogen needs are modest, it can be applied as a sidedressing rather than preplant. Nitrogen needs can certainly be met from organic sources, such as animal manure or a leguminous cover crop. These characteristics enhance pearl millet's desirability in lower-input, dryland production systems. It is advisable to conduct a soil analysis prior to fertilisation.

Irrigation

Little is known about pearl millet's response to irrigation during growth. Irrigation can improve stand establishment if the soil is dry during and after seeding. It appears that pearl millet responds less to irrigation than other grain crops. Greatest water use occurs during the bloom and soft dough stages.

Weed Control

Good weed control is necessary for successful pearl millet production, and it is particularly important to control early emerging weeds. Preventive control options begin with planting clean, weed-free seed. In addition, producers should make sure that all equipment used to plant millet is free of weed seeds. Controlling weeds along ditch banks, roadsides, and field margins will also help prevent weed seed from entering the fields.

Mechanical controls should be used to prepare the seedbed prior to planting millets and where millets are planted in rows for seed, they give producers a headstart on weed control. Because pearl millet is planted relatively late, two preplant tillage operations are recommended, first to stimulate germination of weed seeds, then, several days later, to kill off weed seedlings prior to planting. If planted in wide rows, row cultivation for weed control should be planned, especially if herbicide control is ineffective.

Pest Control

Grain pearl millet is not difficult to grow as it hosts few insect pests. The principal insect problems in millet production are chinch bug, stinkbug, nematode and birds. Normally insecticides are not needed on pearl millet.

Chinch Bug

The female chinch bugs lay eggs, in summer, that hatch into nymphs within two weeks. The nymphs begin to suck the sap from host plants.

The chinch bug inserts its straw-like mouthparts into the plant tissue and sucks out the plant sap while injecting chemicals into the plant which clog the vascular system. The area around the feeding puncture usually turns yellow. Damaged areas first appear as small, irregular patches which enlarge as the insects spread.

Control: Inspect fields every 5 to 7 days until heading. Apply recommended insecticide after heading.

Stinkbugs

Stinkbugs require control on developing grain heads. These insects are most active when pearl millet is planted from July to August. Stinkbug feeding causes small and shrivelled seed. Economic yield loss may occur when 15% or more of the grain heads are infested. Through their needlelike mouthparts, they suck out the plant sap and inject their saliva. As a result, plants become stunted and fruit discoloured. Some plant diseases are also transmitted.

- Control: Stinkbugs can be controlled by the use of registered insecticides.
- Nematodes: Pearl millet hybrids differ in their resistance to nematodes. Primary damage from nematodes is to the roots of the plant. The degree of injury to the roots varies with the age of the plant when attacked. Symptoms are most severe when the feeding occurs during the first few weeks after planting. In general, symptoms consist of greatly reduced root systems with short, stubby roots having dark, shrunken lesions, particularly at the tips. If the root tip is destroyed, new roots may be produced above the damaged area, resulting in a high-rooted appearance. Plants which are not severely damaged by the initial feeding may recover and produce near-normal yields under optimum growing conditions.
- Control: It is recommended to rotate pearl millet with nematode resistant crops.

Bird Damage

Birds readily consume pearl millet seed off the plants in the field. Losses can be severe in small fields or when harvest is delayed for an extended period after maturity.

- Control: Crop monitoring, early planting and timely harvesting are essential to minimise bird damage.

Disease Control

Diseases in pearl millets are not widespread. The most important diseases include mildew, seed rot, rust and grain moulds. Mildew and seed rots can be controlled with recommended fungicide treatment at planting.

Rust can be a major problem of pearl millet. Planting early will minimise yield and grain quality losses owing to rust because rust is a late-season disease. Early planting will allow the crop to mature before disease develops. Fungicide applications are not recommended for pearl millet.

Grain moulds develop when grainfill and maturation occur in wet or humid conditions. Grain moulds may reduce the quality of the grain. It is important to harvest grain as early as possible after maturity, because some grain moulds will increase if harvest is delayed.

Cultural Practices

Diseases in pearl millet can also be controllled by certain sound cultural practices. The following cultural control practices have been found to be economically feasible in reducing disease losses. Growers should properly identify the diseases that limit production and then use a variety of controls in combination.

Rotation with unrelated crops is probably the most utilised cultural practice for disease control. This helps keep populations of pathogens from building up to damaging numbers. One should not expect rotation to eliminate disease development, but it certainly aids in reducing damage from most diseases.

Fertiliser usage may have some bearing on development of certain diseases. It differs with each crop and each disease but, in general, nitrogen out of balance with other nutrients enhances foliar disease development and predisposes some plants to other diseases. Potash, on the other hand, helps reduce disease development when it is in balance with other elements.

Deep burial of crop residue helps to control certain diseases by placing the organism contained in the residue at a depth where there is an oxygen deficiency. This reduces the population of the disease-causing organism and permits the crop to escape much of the damage.

Planting on a raised bed is helpful in preventing certain diseases such as Southern blight and certain of the wilt diseases.

Burning of crop residue has been discouraged because of destruction of valuable organic matter and creation of an air-pollution problem. However, the fact remains, that it is a highly effective means of eradicating some disease-causing organisms associated with crop residue.

Undesirable plants that might serve as a host reservoir for virus diseases, should be removed. Infected rhizomes of Johnson grass are the primary overwintering host for the maize dwarf mosaic virus that attacks grain sorghum, forage sorghum and maize.

Removal of diseased plants as these appear is often an effective method in helping reduce the spread of a destructive disease. Virus diseases of stone fruit and bacterial wilt of cucurbits are examples where roguing is worthy of consideration.

Harvesting

Harvest Maturity

Grain can be harvested as early as 40 days after flowering. When the crop is ripe and dry, grains will pop out cleanly when the head is pinched. You can also begin to harvest when the seed moisture content drops below 15%, but artificial drying to 10 to 12% moisture after harvest is needed to prevent storage moulds. Matured seeds range in colour from white to brown, blue to almost purple. Most are slate grey.

Harvesting Methods

Pearl millet can be harvested by hand, using a knife or sickle. The heads are then dried in heaps on the ground or threshing floor. It can also be harvested by using a combine harvester. Combine harvesting must be adjusted to properly thresh the small seed of pearl millet.

Harvesting for silage is well-suited to mechanical harvest and it should be done when heads appear.

Cultivation Practices of Oilseeds

Groundnut

Groundnut can be grown in both rainy (kharif) and post-rainy (rabi/summer/spring) seasons. The optimum air temperature for growth and development of groundnut is between 25 °C and 30 °C. Moreover, groundnut yield in the rainy season is lower than in the post-rainy season due to cloudy weather and the presence of diseases and insect pests.

Selection of Field and Land Preparation

Groundnut after groundnut in the same field is not advisable as it leads to build-up of diseases and insect pests in the soil. Groundnut should be rotated with a well-fertilized cereal crop. Ideal groundnut soil is well drained, light color with sand, loamy sand or sandy loam texture and pH ranging between 6.0 and 6.3. Maintaining soil pH is important as it affects the availability of nutrients to plants. In case the pH is <5, lime ($CaCO_3$) in appropriate form and quantity should be mixed thoroughly into the soil before land preparation or at the time of land preparation so as to bring it into the optimal range. The rate of application of lime depends upon the type of lime, soil type and depth of application. As a general recommendation, it would require 1.5 t/ha of lime to raise the soil pH from 5.0 to 6.5.

Ridges

Flat beds

The field should be cleared of all stubble and plant residues of the previous crop. Un-decomposed plant residues promote growth of disease causing soil borne fungi. For land preparation, ploughing to a depth of 15-20 cm (very deep ploughing should be avoided) and several passes of the harrow to obtain a fine texture are required. Groundnut can be sown on flat beds or ridges or raised beds separated by furrows. Sowing on raised beds with 0.4-0.8% slope allows easy drainage of excess water, avoids compaction of seed beds and facilitates field operations as all movements are restricted to furrows.

Selection of Variety

Improved, trade-preferred cultivar recommended for the location and cropping season should be selected for sowing. It should also have resistance/ tolerance to diseases and insect pests prevailing in the locality. Some varieties may do well only in a specific growing season.

Quality of Seed

Certified seed purchased from a reliable source or owner saved seed, which is pure (true to type), graded (medium-size), undamaged, fully developed and healthy (free from discoloration and fungal infection) with germination above 90% should be used. Germination test on seeds should be carried out one week before sowing and the seed rate should be adjusted accordingly. Gap filling after 8-10 days of sowing does not help much.

Cultural Practices

Manure and Fertilizers

Groundnut responds to residual soil fertility better than the direct application of fertilizers. The crop(s) preceding groundnut should be well fertilized to build up soil fertility particularly for phosphorus (P) and potassium (K). Application of fertilizers and their dose should be based on the nutrient status of the soil as determined by the soil test and the targeted yield. However, general recommendations for groundnut are as follows:

Manure

Farm yard manure (FYM) or Compost: 10-12 t/ha; 25-30 days before sowing. Introducing green manure in crop rotation also helps to increase the organic matter content of the soil and improve its structure.

Macronutrients

Nitrogen (N), P and K: 8-20 N, 16-80 P_2O_5, 0-75 K_2O kg/ha; as basal application. Calcium (Ca): 200-400 kg/ha of gypsum at the peak flowering stage as side placement. Calcium is essential for good seed development.

Micronutrients

Many fields are deficient in micronutrients – boron, zinc and sulphur. If soil test shows deficiency of these micronutrients, remedial measures should be taken as follows:

- Boron (B): Apply 3-4 kg/ha borax to the soil at the time of land preparation. The residual effect of borax should last several seasons. Alternately, 0.1% borax can be sprayed on the crop early in the season to ensure boron uptake before flowering.
- Zinc (Zn): Apply 10-20 kg/ha zinc sulphate to the soil once in three years at the time of land preparation.
- Sulphur (S): Application of gypsum provides adequate sulphur to the crop. Sulphur deficiency is most likely on very sandy soils, which possess little anion exchange capacity.
- Iron (Fe): In many calcareous soils, groundnut plants show iron deficiency symptoms (interveinal chlorosis in young leaves followed by full chlorosis (whitish yellow) of entire leaves). Iron chlorosis can be alleviated by applying ferrous sulphate @ 10 kg/ha to the soil or spraying the affected crop with 0.5% ferrous sulphate + 0.2% urea solution. If required, the spray treatment could be repeated at 10-14 days interval.

Zinc deficiency

Sulphur deficiency

Iron deficiency

Seed Treatments

Rhizobium inoculation: Rhizobium inoculation could be beneficial in newly cleared fields, rice fallows, fields with eroded soils and low fertility. Seeds should be treated just before sowing with Rhizobium culture.

For soil borne diseases: Seeds should be treated with captan (1.5 g) + thiram (1.5 g), carbendazim (2.0 g) or mancozeb (3.0 g) per kg of seed or other locally recommended fungicide(s). Seed treatment with Trichoderma viride or T. harzianum @ 4-5 g/kg seed also helps in managing seed and soil borne diseases.

For soil insect pests: In white grub and termite endemic areas, seed should be treated with chlorpyriphos 20 EC @ 12.5 ml/kg seed.

For sucking insect pests: Seed treatment with imidacloprid (17.8 SC) @ 2 ml/ kg seed gives protection against sucking insect pests (thrips, jassids and aphids) and leaf miner at early stages of plant growth.

Seed dormancy: Virginia varieties have postharvest seed dormancy, which may last for 5-6 months. If such varieties are to be sown immediately after harvest, the seeds should be thinly spread over a tarpaulin or plastic sheet and sprayed thoroughly with etherel 39 EC @ 5 ml/L water and air dried just before sowing.

Seed treatment with fungicide/insecticide gives protection up to 30 days after sowing. It should be carried out one or two days prior to sowing. Seeds should be treated first with liquid chemicals and after drying with powder/dust chemicals. If the Rhizobium strain is not compatible with fungicide/insecticide, the culture can be applied in sowing rows following slurry method.

Spacing, Sowing Depth and Seed Rate

- Spacing: Row to row 30–45 cm and plant to plant within a row 10–15 cm or as recommended for the location; one seed/hill.
- Sowing depth: 5 cm
- Optimum plant population: 330,000 plants/ha for Spanish/Valencia cultivars and 148,000 plants/ha for Virginia cultivars.
- Seed rate: The seed rate will vary depending on seed weight, germination % and row to row and seed to seed spacing adopted. Normally, it may range from 100 kg to 160 kg/ha.

Sowing Method

The use of seed drill (bullock-drawn or tractor-mounted) is recommended as it results in faster sowing, quicker emergence and uniform plant stand. There should be enough moisture in the soil before sowing to ensure quick and uniform germination. In case of dry sowing, irrigation should be provided soon after, preferably with sprinklers. Flood irrigation should be avoided.

Intercultivation and Weed Management

It is essential to keep groundnut fields weed free for up to 45 days after crop emergence. Even at later stages it is desirable not to have weeds in the field as they interfere with harvesting. Application of pre-emergence herbicides such as pendimethalin @ 1.0 -1.5 kg a.i./ha as spray or fluchloralin @ 1.0 -1.5 kg a.i./ha as pre-plant soil incorporation followed by 1-2 hand weeding, as and when needed, effectively reduces weed competition. The last hand weeding can be done along with gypsum application so as to incorporate it in the soil. The plant should not be disturbed once the pegs enter the soil. Interculture in a rainfed crop helps to reduce weeds and also encourages infiltration of rainwater.

Many farmers practice earthing up (mounting soil around the plant), which is not recommendable, to allow pegs from higher nodes to enter the soil. This practice may promote growth of stem rot causing fungus (Sclerotium rolfsii). It also deteriorates the quality of earlier set mature pods while waiting for the later set pods to mature.

Water Management

- Rainfed crop: Proper arrangements for drainage should be made so that excess rain water does not stagnate in the field. If supplementary irrigation is available, it should be given at critical stages such as flowering, pegging and pod and seed development.

- Irrigated crop: Except for crops grown on residual moisture, rabi/summer/spring season crops are fully irrigated. Generally, 600-650 mm water is sufficient to raise a full groundnut crop. A 2-3 week moisture stress soon after crop emergence followed by regular irrigation, often helps in inducing profuse flowering and uniform pod maturity. At pegging and pod and seed development stages, light but frequent irrigation is required. Excessive irrigation at later stages of crop growth may promote pod and seed diseases at maturity. The preferred method of irrigation is sprinkler irrigation. Flood irrigation, often practiced in flat sowing in south Asia, is not a good method of irrigation as it wastes water, results in over watering and trampling of plants in the field by workers engaged in irrigation.

Plant Protection

There are a number of disease/insect pest management measures, including use of resistant cultivars, cultural, chemical and biological. Growing resistant/tolerant cultivars is the most economic and efficient measure. In case the level of resistance in a preferred cultivar is not high enough, other approaches should be combined to obtain better protection against diseases and insect pests.

For high volume sprayers, 450-500 L and for low volume sprayers 225-250 L water is required to cover 1 ha. While using chemicals, protective clothing should be worn and proper care should be taken to dispose empty bottles/cartons of chemicals in a safe manner.

Diseases

Rust and early-and Late-leaf Spots

If both rust and leaf spots occur together, chlorothalonil 75 WP @ 750 g a.i./ ha should be sprayed on the crop. If only leaf spots appear, use carbendazim 50 WP @ 250 g a.i./ ha or mancozeb 50 WP @ 500 g a.i./ha. If only rust is seen, use calixin 80 EC @ 250 ml a.i./ha. For effective control, fungicides should be applied immediately after the appearance of symptoms. Further applications should be made at 10-15 days interval until 2-3 weeks before harvest. A close monitoring of the crop can help to reduce the number of sprays to the minimum.

Collar Rot/Stem Rot

These diseases can be managed by treating seeds with seed dressing fungicides or T. viride or T. harzianum or soil application of T. viride or T. harzianum @ 2.5 kg (incorporated in 50 kg FYM or castor cake) per ha at the time of sowing.

Aflatoxin

Aflatoxin, a carcigenic toxin, is produced in groundnut seed by Aspergillus group of fungi. The fungus can infect the pods in the field when they are developing, during harvesting and curing and in storage. End-of-season drought influences pods to fungal infection. It is essential to minimize aflatoxin contamination of the produce as it affects human and livestock health and restricts the international trade of the commodity. The following precautions can be taken to minimize aflatoxin contamination growing tolerant varieties, applying Trichoderma viride @ 1 kg (mixed with 50 kg FYM) per ha to the soil at the time of sowing and gypsum at peak flowering, light but frequent irrigation (if available) during pod and seed development stages, avoiding mechanical damage to pods during weeding, harvesting, curing, threshing and storage, control of soil insects, harvesting at optimum maturity (in case of severe drought, the crop should be harvested early), drying the pods to <8% moisture content, removal of immature, discolored and damaged pods from the produce, not mixing the gleanings (leftover pods collected from the soil) with main produce and protection from storage insect pests. Immature pods should be removed from the haulms before feeding them to livestock.

Peanut Bud Necrosis Disease and Peanut Stem Necrosis Disease

Seed treatment with imidacloprid provides initial protection for 30 days after sowing through control of thrips vectors. Field sanitation, removal of weed hosts and cultural practices such as timely sowing, optimum plant population and intercropping and border cropping with fast growing tall cereal crops offer best solution to contain these diseases.

Insect Pests

Tobacco Caterpillar (Spodoptera), Gram Pod borer (Helicoverpa) and Red Hairy Caterpillar (Amsacta). These defoliators inflict economic losses only when the foliage damage exceeds 25%.

Tobacco caterpillar | Gram pod borer | Red hairy caterpil-lar

- Integrated pest management (IPM): Includes growing of resistant/tolerant cultivar, growing of trap crop such as sunflower and castor bean on borders or in groundnut field (1 plant 20/m), destroying of egg masses on trap crops and groundnut plants by hand, encouraging larvae predation by birds by providing perches in the field (10-15/ha) and application of nuclear polyhedrosis virus (NPV) @ 250 LE/ha, obtained from a reliable source and neem seed kernel extract @ extract obtained from 10 kg neem seed powder/ha.
- Chemical control: Chemical control should be followed as a last resort when there is absolute necessity. Apply endosulfan @ 350 mL a.i./ha, or monocrotophos @ 300 mL a.i./ha or fenvalerate @ 100 mL a.i./ha or indoxacarb @ 70 mL a.i./ha or spinosad @ 45 mL a.i./haif defoliation exceeds 25%, or if one or more larvae per plant is observed during the first 50 days after seedling emergence (DAE).

For red hairy caterpillars, digging a 15-20 cm deep trench all around the field and placing Ipomea or some other plant twigs in them can restrict the migrating caterpillars to the trench, where they could be destroyed manually or by spraying fenvalerate @ 1 ml/L water. Alternately, a short barricade of polythene fence (10 cm high) across the migrating route can prevent their entry into the field and they can be collected manually and destroyed.

Groundnut Leaf Miner

Chemical control is recommended if 5 or more active larvae/plant are found up to 30 DAE, 10 larvae/plant between 30-50 DAE, or 15 larvae/plant at 51 DAE or later. Apply dimethoate @ 200-250 ml a.i./ha or monochrotophos @ 150-200 ml a.i./ha or imidacloprid @ 25 ml a.i./ha.

Aphids, Thrips and Jassids

For aphids, apply dimethoate @ 200-250 ml a.i./ha when all the terminal buds in a young plant are infested. Only when five thrips per terminal leaf (folded) are observed before 20 DAE, apply imidacloprid @ 25 ml a.i./ha or dimethoate @ 200-250 ml a.i./ha. If 10% of the total leaves show hopper burn ('V' shaped yellowing at the leaf tip), imidacloprid @ 25 ml a.i./ha or dimethoate @ 100- 200 ml a.i./ha should be sprayed for control of jassids.

Aphids Thrips Jassid

White Grub

Apply thimet 10 G or carbofuran 3 G @ 1 kg a.i./ha in seed furrows just before sowing or treat seeds with chlorpyriphos 20 EC @ 12.5 ml/kg seed.

Termites

Seed treatment with chlorpyriphos provides protection up to 30 days after sowing. Chemical control in rainfed crop at later stages is difficult and expensive. However, some cultural practices such as destroying the termite mounds in the vicinity of the field, removal of plant residues and debris from the field and timely harvest can help to minimize the damage.

Harvest and Postharvest Operations

Harvest at Maturity

Optimum harvest time is determined by uprooting a few representative plants from different spots in the field around the expected time of maturity and checking the inner side of the pod shell, which shows black color when mature. When 75-80% pods in case of Spanish/Valencia cultivars and 70-75% pods in case of Virginia cultivars show internal pericarp darkening, the crop is ready for harvest. If sprouting of seeds is observed in Spanish/Valencia cultivars (due to rains at harvest time in cultivars lacking fresh seed dormancy), the crop should be harvested as soon as the conditions permit without waiting for 75-80% of pods to mature. Over maturity or delay in harvesting can result in greater pod loss in the soil and deterioration in pod quality.

Drying and Curing

Lift the plants, and invert them with the pods uppermost in windrows for about 2-3 days. Pick the pods (preferably by thresher) and spread them out in a thin layer to sun-dry for a further 3-4 days. The seeds in well-dried pods should have less than 10% moisture content. In the post-rainy season, when temperatures at the time of harvesting are high, the harvested plants should be assembled in circular heaps with pods facing inside so as to avoid their direct exposure to the sun. The pods should be shade-dried to maintain seed viability.

Inverted windrows

Circular heaps

Storage

It is important to remove all damaged, discolored, rotted, immature and sprouted pods, other plant materials and soil from the produce before storage. Under unfavorable conditions, groundnut seed loses viability quickly.

Commercial Produce

Well cleaned, dried, mature pods free from plant debris, soil and other inert materials should be stored in gunny bags in a well-ventilated rodent free room for marketing. The bags should be placed on wooden planks and should not be stacked very high. If longer storage is needed for marketing reasons, proper care should be taken to avoid storage insect pest infestation. Only safe chemicals, if needed, should be used.

Mustard

Mustard is an annual herb with seedlings that emerge rapidly, but then usually grow slowly. Plants cover the ground in 4 to 5 weeks with favorable moisture and temperature conditions. The tap roots will grow 5 ft into the soil under dry conditions, which allows for efficient use of stored soil moisture. Plant height at maturity varies from 30 to 45 in. depending on type, variety, and environmental conditions.

Flower buds are visible about five weeks after emergence. Yellow flowers begin to appear 7 to 10 days later and continue blooming for a longer period with an adequate water supply. A longer flowering period increases the yield potential. About half of the flowers produce dark, reddish-brown seeds that are retained in pods of 0.5 to 0.75 in. in length. Flowers pollinated during the first 15 days of the flowering period produce most of the seed.

Environment Requirements

Climate

Mustard is a cool season crop that can be grown in a short growing season. Varieties of yellow mustard usually mature in 80 to 85 days whereas brown and oriental types require 90 to 95 days. Seedlings are usually somewhat tolerant to mild frosts after emergence, but severe frosts can destroy the crop. Mustard, especially the brown and oriental types, has a partial drought tolerance between that of wheat and rapeseed. Moisture stress caused by hot, dry conditions during the flowering period frequently causes lower yields.

Soil

Mustard can be raised on variable soil types with good drainage, but is best adapted to fertile, well-drained, loamy soils. Soils prone to crusting prior to seedling emergence can cause problems. This crop will not tolerate waterlogged soils since growth will be stunted. Dry sand and dry, sandy loam soils should also be avoided.

Seed Germination

Seed will germinate at a soil temperature as low as 40 °F.

Cultural Practices

A small grain crop following mustard in the rotation will usually yield more than when following continuous small grain. Mustard has several of the same diseases and insect pests as flax, oilseed rape (canola), sweet clover, soybeans, field peas, lentils, and sunflowers. Therefore, crops from this group should be avoided in the same rotation as mustard. Cereal grains are not very susceptible to the pest and disease problems of mustard.

Seedbed Preparation

The seedbed should be firm, fairly level, and free of weeds and previous crop residue.

Soil is firm enough for seeding when only a shallow depression of a heel is made when someone stands on the soil. Shallow tillage, just deep enough to kill weeds, should keep soil moisture close to the surface and leave a firm seedbed. If necessary, the seedbed should be packed before planting to obtain a firm seedbed. Firm seedbeds with adequate moisture allow shallow planting and encourage rapid, uniform seed germination and emergence of seedlings.

Seeding Date

Planting should occur as early in the season as the environmental conditions allow. The soil temperature should be at least 40 to 45 °F at a depth of 1 in. If seedlings are damaged by frost after emergence, 4 or 5 days may pass before the full extent of the damage is known. Plants should recover if the growing points are not destroyed. An earlier seeding date allows plants to benefit from the spring moisture in establishing a good canopy before weeds emerge, and to avoid heat stress during summer that causes flower or pod abortion. Early seeding also reduces the risk of damage from fall frosts that can reduce crop yields and quality. The recommended seeding date in northern Minnesota and Wisconsin is May 1 to 25. Seeding later than May 15 frequently results in lower yields.

Method and Rate of Seeding

Yellow mustard, which has approximately 100,000 seeds/lb, is solid seeded with a grain drill at a rate of 8 to 14 lbs/acre. The higher rate should be used on heavy, fertile soils or on those where emergence is difficult. Brown and oriental mustards have 200,000 seeds/lb and should be solid seeded at a rate of 5 to 7 lbs/acre. Seed is small and must be planted shallow at a 1/2 to 1 in. depth. If very dry soil conditions exist seeding depth should be increased to 1 1/2 in. If mustard stands are poor, quick decisions for reworking and reseeding should be made.

Fertility Requirements

Mustard generally responds to nutrient additions in a similar way as does rape or canola. Soil tests should be used to determine nutrient need. Optimal (medium) soil test levels are about 15 to 20 ppm Bray P, and 80 to 100 ppm K. At these levels fertilizer should be applied at a rate of about 45 lbs/acre P_2O_5 and 80 lb/acre K_2O. When fertilizer is banded, the bands should be placed below and to the side of the seed furrow. Mustard responds well to nitrogen additions with optimum yields occurring at about 100 to 120 lb/acre N. Where mustard follows legumes or manure additions, appropriate credits should be taken.

Work in the western U.S. shows that mustard responds well to sulfur on low S-supplying soils. Sulfur fertil-sandy soils in northwestern Wisconsin and northern Minnesota which have not been manured in the past two years.

On soils deficient in boron (testing at less than 0.5 ppm B), apply 0.5 to 1 lb/acre in a uniform broadcast application. Never band B near the seed. Soils with a pH near neutral (7.0) are desired for this crop. Nevertheless, an alkaline pH and slightly saline soils are tolerated. Mustard has a tolerance to soil salinity that is similar to barley.

Weed Control

Weeds can greatly reduce mustard yields. Weed seeds, which are difficult to remove, can cause high losses during seed cleaning and lower market grades. Good weed control is based on preparation of a clean field and shallow seeding to encourage quick, uniform emergence. Young mustard seedlings do not compete well with weeds and the early establishment of a uniform, vigorous crop helps control annual weeds. The crop cannot be cultivated after emergence.

Mustard, especially the oriental and brown types, should be grown on land with as little wild mustard as possible to avoid costs of removal and loss of tame mustard seeds. Wild mustard seed can be mechanically separated from the larger-seeded yellow type, but separation is not possible with the smaller-seeded brown and oriental types. Wild mustard seed often reduces the crop quality to the sample grade. Production of rapeseed and mustard on the same fields is also not recommended since seed mixtures can occur easily and degrade both crops.

Control of perennial weeds such as Canada thistle, field bindweed, and quackgrass, should be started in the fall or prior to planting in the spring. Control Canada thistle by applying Roundup (glyphosate) before the last killing frost in the fall when it is still actively growing. Spring treatment of Canada thistle prior to planting is not usually adequate for complete control. Apply Roundup to quackgrass when it is at least 8 in. tall and growing actively. Allow 3 days between application and tillage. Control field bindweed with Roundup when it is actively growing on moist sod and is at or past full bloom, preferably in late summer or fall the year before planting mustard. Spring treatment to control perennial weeds before planting is usually not practical.

Trifluralin (Treflan EC) is labeled for use on mustard in Minnesota and North Dakota for control of a wide variety of grasses and broadleaf weeds; however, it will not control wild mustard. The rate of herbicide applied will vary with the soil type, organic matter content, and species of weeds that need to be controlled. Check the label for the correct rate to use on your fields. Trifluralin must be applied before seeding and incorporated thoroughly in the soil for maximum effectiveness. Mustard is sensitive to the broadleaf herbicides used on cereal crops, such as 2, 4-D and MCPA, and spray drift from adjacent fields must be avoided. Crops that can be sprayed with 2, 4-D or MCPA should follow mustard in the rotation so volunteer plants can be controlled.

Diseases and Control

This crop is vulnerable to several diseases, among which the most serious are Sclerotinia stalk rot (white mold), downy mildew, white rust, leaf spots, and mosaic virus. Good cultural practices are the most effective control measures for diseases. These practices should include keeping records of disease occurrence, compliance with the proper crop rotation, control of host plants for diseases in fallow fields and non-crop areas, and use of seed treatments.

Sunflower, rapeseed, canola, safflower, soybeans, crambe, and drybeans should not be grown in rotation with mustard since they have similar disease problems. If these crops are produced in rotation with mustard, damage from these diseases can increase to economic levels. Numerous broadleaf weeds can also serve as hosts or sources of infection for these diseases. Wild mustard, pigweed, field pennycress, and shepherd's purse are examples of predominant hosts. One of the best methods to avoid serious disease problems in mustard (and leaf diseases of small grains) is to produce this crop in a small grain rotation. Mustard should be spaced four or more years apart in the crop rotation to avoid problems with soil-borne diseases. After potato or flax crops, one year should pass before mustard is raised on the same field due to the presence of root rot or damping-off pathogens.

Insects and other Pests

Growers should monitor fields closely to detect insect problems that can result in significant yield losses. Flea beetles and caterpillars of the diamondback moth have been the most serious pests. Young seedlings can be seriously damaged by flea beetles soon after emergence. Feeding activity of adult beetles causes a shot-holed appearance to cotyledons and the first true leaves. Damaged plants may die or suffer a reduction in vigor. Hot, sunny weather encourages feeding while cool, damp conditions slows insect feeding and promotes crop growth. Injured plants may wilt and die during hot, dry weather, which results in mild to severe yield losses. Serious crop damage does not usually occur once the crop develops beyond the seedling stage since vigorous plants can outgrow beetle defoliation.

Losses caused by flea beetles can be minimized by practicing the proper cultural methods. A well-tilled, firm seedbed with adequate fertility should permit young seedlings to outgrow beetle damage during the vulnerable stages early in the season. The presence of a few flea beetles or scattered shot-holing is not cause for serious concern. However, if beetles are numerous and feeding damage is present on most cotyledons, prompt control may be necessary. Malathion EC at 1 1/4 lbs/acre, Carbaryl (Sevin) at 1 lb/acre, Ethyl parathion 8E at 1/2 lb/acre, and Thiodan EC at 3/4 lb/acre can be used for control of flea beetles. Read labels for waiting period and correct timing.

Larvae of diamondback moths eat leaves, flowers, and green seed pods. Malathion used at 2 1/2 lbs/acre will control these caterpillars. Damage from sugar beet nematodes can be avoided when sugar beets are not grown on the same field two years before or after mustard, since mustard is a host plant for this nematode.

Harvesting

The normal maturation of the crop, wind, and rain do not cause shattering before cutting. However, the actual harvesting operations can cause great shattering losses when the plants are overripe. Yellow mustard does not shatter readily and can be straight combined if the crop has matured uniformly (10% moisture) and is free of green weeds. If the crop is weedy or uneven in maturity it should be swathed. Swathing, if deemed necessary or preferred, should be done when 60 to 70% of the seed has turned yellow-green. Plants should be cut just beneath the height of the lowest seed pods. The swath will then settle into the stubble and reduce the chance of being blown by high winds. Yellow mustard does not cure quickly. Straight combining is therefore recommended at 12 to 13% moisture, followed by artificial drying, to obtain uniform quality and highest yield.

Brown and oriental varieties will shatter more readily when ripe and should be swathed. The swathing should begin after the general leaf drop and when the overall field color has changed from green to yellow or brown. Pods sampled from the middle of racemes from several plants, in arm representing the average maturity, should be examined for physiological maturity. About 75% of the seeds may have reached the mature color of yellow or brown. The remaining green seeds will mature in the swath before combining.

Swathing should be done under conditions of high humidity or when morning dew is on ripe pods to decrease shattering losses. Windrows tend to be bulky and subject to scattering by the wind. A roller or steel drum should be used to press the swath into the stubble. The combine should be adjusted so seeds are threshed completely by using the lowest cylinder speed, which is set at approximately 600 RPM, and the appropriate cylinder opening. The reel may cause shattering when straight combining, but it can be removed or lifted above the plants it the stand is good. If the reel is needed, remove half of the bats and reduce its speed. Cylinder speed may need to be adjusted during the day as crop moisture content may vary.

Drying and Storage

When the mustard seed reaches a moisture content of 10% or less it can be stored safely. The harvested seed should be handled carefully since it will crack easily when moved in and out of storage. The damaged seed becomes dockage and is a loss to the grower. Air temperatures for seed drying should not exceed 150°F and seed temperature should stay below 120°F. Use of drying equipment designed for corn or wheat may require some modification when drying mustard. A fine screen will be needed to prevent loss of the smaller seed. Storage bins must be free of cracks or holes.

Cultivation Practices of Pulses

Chickpea

Land and its Prwparation Land preparation for sowing chickpea is based on the soil type and cropping system. In the case of a heavy soil, a rough seedbed is prepared to avoid packing of the cloddy surface due to winter rains and to facilitate soil aeration and easy seedling emergence. When chickpea is cultivated as a mixed crop with linseed or mustard, the land is prepared to a fine tilth. It is necessary to deep-plow the field at the beginning of the rainy season. This opens the soil deep and ensures efficient moisture conservation. Deep plowing also reduces wilting of chickpea that tends to develop due to the presence of hardpans in the root zone.

- Seed rate (Kg/ha.) - Seed rate of 75-100 kg per hectare depending upon seed size may be sufficient for one hectare. The seed should be placed 8-10 cm deep.
- Nutrient Management - Chick pea being a leguminous crop fulfills the major part of its nitrogen requirement (about 75%) through the process of symbolic nitrogen fixation which works effectively from three to four weeks after sowing. However, soils with low organic matter and poor nitrogen supply may require 20-25 kg per hectare of nitrogen as starter does which can meet plant requirement before the formation of nodules. Besides nitrogen, pulses respond very favourably to phosphorous application if the soils are deficient in phosphorous supply. If both nitrogen and phosphorous are required to be supplied then diammonium phosphate (18-46-0) at the rate of 100 to 150 kg per hectare should be applied uniformly before the last discing ploughing.
- Water Management - Chick pea is mostly sown as a rainfed crop. However, where irrigation facilities are available, give a pre-sowing irrigation. It will ensure proper germination and smooth crop growth. If winter rains fail, give one irrigation at pre-flowering stage and one at pod development stage. In no case first irrigation should be given at flowering time of gram crop. A light irrigation should be given because heavy irrigation is always harmful to gram crop. Excess of irrigation enhances vegetative growth and depresses chick pea yield.
- Weed Management - Chick pea being a stature crop suffers severely by infestation of weeds. One hand weeding or inter culture with hand hoe or wheel hoe after

25-30 days and second if needed after 60 days of sowing may take care of weeds. Fluchloralin (Basalin) 1 kg per hectare in 800-1000 liters of water as pre-planting spray may be used as an effective herbicide. It should be well incorporated in the soil before sowing. In case Basalin is not available use Metribuzin or Prometrynen at the rate of 1.0-1.5 kg active ingredient in 800- 1000 liters of water per hectare as pre-emergence spray. Hand weeding or inter culture with the help of hoe is always better than herbicides because inter culture operations improve aeration in the soil.

Diseases

- Wilt (Fusarium orthocerus)- Treat the seed with Benlate T or a mixture of Benlate of Thiram (1:1) at the rate of 2.5 g per kg of seed.
- Sclerotinia Blight (Scleritinia sclerotiorum)- Use only healthy seeds free from sclerotia. After harvest, the diseased plants should not be allowed to stand in the field but should be destroyed by burning. Treat the soil with a mixture of fungicides like Brassicol and Captan at the rate of 10 kg per hectare.
- Rust (Uromyces ciceris arietini)- With the appearance of first symptoms, spray the crop with 0.2% Mancozeb 75 WP followed by two more sprays at 10 days interval.

Insect Pests

- Gram Pod Borer (Helicoverpa armigera)- Spraying of Spinosad 45SC @ 0.1ml/lit or Indoxacarb 14.5SC @ 0.3ml/lit or Flubendiamide 48SC 0.1ml/lit or Novacuron 10EC 1ml/lit of water.

Crop becomes ready for harvest when leaves turn reddish-brown and start shedding. Plants are either plucked out by hand or cut with sickle. The crop is allowed to dry in sun on threshing floor for about five to six days. Thereafter, threshing is done either by beating the plants with sticks or by trampling under the feet of bullocks.

Lentil

Lentil is a small, green annual herb with branched stem. It is rarely tall than 40 cm. The pod is flattened and broad about 12 mm long, tipped by a minute beak. The seeds are lens-shaped, light red, brown, grey and of tan speckled with black. Lentil is recognized as one of the most nutritious pulse crops, ranking next to chickpea (gram) amongst rabi pulses.

Soil and Climate

Lentil is grown on a wide range of soils ranging from light loamy sand to heavy clay soils.Its range of cultivation extends to an altitude of 3500m. The optimum temperature for its growth and development ranges from 15 to 25 °C.

Land Preparation

After harvest of kharif crop, land should be ploughed once with soil-turning plough. This should be followed by cross-harrowing and planking. Raised bed planting is beneficial than flat bed sowing.

Cropping Systems

It is generally grown as rainfed crop during rabi after rice, maize and pearl millet. In intercropping it is grown in barley, linseed and mustard. It is also grown as an intercrop in autum-planted sugarcane.

Seed Rate

The seed rate for small-seeded varieties is 30-40 kg/ha and for bold seeded varieties, it is 50 – 60 kg/ha.

Sowing

The seed is sown in the second forthnight of October under rainfed conditions. However, it can be sown in the first week of November in irrigated areas. Under late sown conditions, the seed can be sown upto second week of December.

Spacing

The plant to plant distance should be 10 cm and row-to-row spacing is 25 cm.

Nutrient Requirement

The recommended dose of fertilizers is 20kg N, 40kg P, 20 kg K and 20kg S/ha. In soils low in Zinc, soil application of 20 kg $ZnSO_4$ is recommended under rainfed and and late sown conditions. Foliar spray of 2% urea improves yield.

Water Management

In the absence of winter rains, light irrigation at 40-45 days after sowing and at pod-formation stage is beneficial.

Weed Management

One weeding 30-40 days after sowing or use of weedicide (Pendimethalin @1.0 kg/ha) immediately after sowing helps in controlling weeds.

Pest and Disease Management

The major diseases of lentil are rust and wilt. Use of resistant varieties is helpful in controlling the disease. Pests are not a major problem in lentil. In case of aphid infestation, spraying of Monocrotophos (0.04%) is effective.

Harvesting

When more than 80% pod matures, it should be cut with a sickle, dried on the threshing floor and threshed by beating with sticks or trampling with bullocks.

Yield

The average yield is 8 to 10q/ha under rainfed and 12-15q/ha under irrigated conditions. Under good management, the yield may go up to 20q/ha.

Cultivation Practices of Fibre Crops

Cotton

Cotton is one of the most important fiber crop and plays a dominant role in the industrial and agricultural economy of the country. cotton is known for its intensive cultivation. Some production practices like wide plant to plant and row to row spacing and crop traits such as indeterminate growth habit, long duration, render the crop susceptible to a multitude of pests and diseases at all stages of growth. Aggressive production practices by farmers often lead to a very high input use, with little regard to matching returns. The excessive use of inputs, not only escalates the cost of cultivation but also decreases the profitability. In order to address these issues, a cultivation practice has developed the concept of operationalizing Better Management Practices (BMPs) for cotton cultivation.

Better Management Practices, are more environment friendly, and promote the use of local resources, and improve the input use efficiency. The practices, apart from improved agronomic practices, can be broadly categorized into five broad areas, namely:

- Soil Fertility Management.
- Water Management.
- Pest and Disease Management.
- Proper Harvesting and Storage.
- Crop Residue Management.

Soil Management

Soil Nutrition

Most of the cotton growing tracts in the country are characterized by low to medium levels of available Nitrogen (N) and Phosphorous (P) and medium to high levels of available Potassium (K). Organic matter content varies between 0.3 – 0.56%.the fertility index of cotton growing soils is not very encouraging. Furthermore, the extraction of nutrients is enormous as cotton is a deep rooted crop. Again the nutrient uptake varies with the soil type. Cotton is generally grown in vertisols of varying depth (shallow, medium deep and deep soil) in the rainfed regions of the central zone. More than 67% of the cotton growing areas fall under shallow and medium deep soil. The nutrient uptake of cotton in different soil types is given below:

Soil Type	N uptake (kg/ha)	P uptake (kg/ha)	K uptake (kg/ha)
Deep	55	20	57
Medium deep	44	13	43
Shallow	44	12	45

It is therefore important to replenish the soil nutrients through balanced fertilization for long term sustainability. In order to enhance the nutrient use efficiency it is important that the nutrients are applied at the right times, in right quantities and using appropriate methods.

Nutrient	Method of application
Nitrogen	The recommended dose of N fertilizer as split into three applications:1/3rd quantity each at 30, 60 and 90-days after sowing
P & K	Entire recommended dose of P fertilizer to be applied as a basal application at the time of last plough. Recommended dose of K split into three applications: 1/3rd quantity each at 30, 60 and 90 days after sowing

The recommended doses for a given area and under different farming systems have to be ascertained from the local agricultural research stations.

Soil Characteristics

Apart from the nutritional capacity of the soil, there are other aspects which need serious attention:

- Soil Moisture: For the soil to be healthy, it should have good water holding capacity. For the rainwater to percolate into the soil without wastage through run off, the soil should be loose and friable. In addition to the soil microbial organisms, organic matter content helps in holding water.

- Soil Resilience: The ability of the soil to withstand various natural processes and management interventions and still be amenable for sustainable crop production is known as its resilience. Soil needs to be protected from wind and water erosion. Management practices that help build up the soil organic matter and microbial activity are important to safe guard the productivity of soils. Ensuring proper land use and land cover through appropriate measures such as mulching, cover cropping and crop residue recycling improve its regenerative capacity.

Identification of Plant Nutrient Deficiency Symptoms

All nutrients play an important role in the metabolic activities of a plant. A nutrient higher in quantities cannot replace the role of another deficient nutrient. Plants express deficiency of a nutrient through symptoms on their leaves, stems and in their growth.

Some nutrients are highly mobile, some less mobile while some immobile in the plant system after uptake by the plant through its roots from the soil.

Highly mobile nutrients include Nitrogen, Phosphorous, Potassium, Magnesium and Chloride. When the soil becomes deficient of these nutrients, the deficiency symptoms first appear on the lower/older leaves of the plant as these nutrients get translocated to the younger leaves/apical growth because of their mobility in the plant system.

Immobile nutrients are Iron, Boron, Copper and Calcium. When the soil is deficient in these nutrients, they cannot move towards apical parts of the plants as they are immobile, the deficiency symptoms appear first on the apical parts of the plant (upper/younger leaves).

Nutrients with limited mobility include: Zinc, Sulphur, Manganese and Molybdenum:

- N Deficiency: Early season deficiency results in the plant with pale green yellowish leaves and stunted growth and late season deficiency leads to reduced boll retention.
- P Deficiency: Reflected in the lower leaves, plants are stunted and have dark green leaves, sometimes impart a purplish yellow color to the leaf.
- K Deficiency: Reflected in the lower leaves, occurs as interveinal chlorosis, dry leaves and premature shedding.
- S Deficiency: Seen on upper young leaves, first turn to light green, then to light yellow followed by a pronounced yellowing.
- Ca Deficiency: Manifests as deformed and chlorotic leaves at the growing tip and stunted growth.

Soil Test based Nutrient Application

Soil should be tested for its chemical properties by using the portable soil analysis kit. The analysis report can be used to plan the fertilizer application. Micro nutrients deficiency is made up with proper application of compost/manure along with the main fertilizer. Cotton is generally prone to secondary nutrient deficiency particularly Mg. In case of visible deficiency symptoms, Mg and other micro nutrients are applied as foliar sprays at appropriate growth stages.

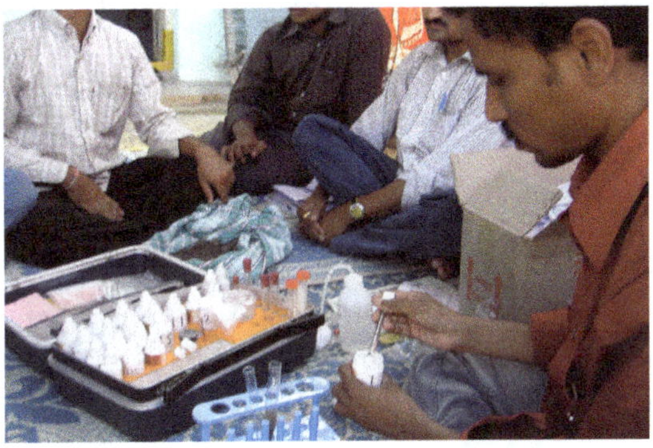

Soil testing kit being used for field soil testing

Organic Practices

Better Management Practices have been developed in such a way that they require the use of organic input practices in conjunction with inorganic fertilizers.

Different practices are mentioned below so that farmers can use these simple techniques and enhance the soil organic matter.

- Tank Silt Application: Tanks are the naturally formed depressions (small lakes) with a strengthened bund.They normally have catchments that are rich in nutrients. Soil loss in the catchments along with run off results in sedimentation of fertile top soil in the tank bed. The silt, which is finer fraction of the soil deposited in the tanks, is excavated and applied as an external amendment in cotton fields to improve water retention capacity and fertility.

Tank Silt Application

- Vermicompost: Non-burrowing earthworms can convert organic crop and animal residues into compost which is rich in both major and minor nutrients and other plant growth promoting substances. Farmers are aware on the use of vermicompost.

The project supports the farmers to produce vermicompost by constructing simple, low cost vermicompost beds. Cement rings (3-4 feet diameter) are prepared at the village level as a group activity, which reduces the cost of construction. These cement rings are then used in constructing the compost beds. Farmers are encouraged to use 2-3 quintals of vermicompost per acre, as per the availability.

Preparation of vermicompost

- Sheep penning, cattle penning and poultry manure: Farmers are encouraged to use sheep penning and poultry manure. Sheep penning and poultry manure are used in the cotton fields, where tank silt can not be applied. This is suggested to reduce the cost of cultivation and to use resources efficiently.

Cattle penning

- FYM Application: Farm Yard Manure (FYM) is generally applied in cotton fields. Farmers understand the importance of FYM application and feel the need for it; therefore additional efforts are not required to motivate them. However, there is a greater need to motivate the farmers to change the technique of FYM preparation, as the process is prone to nutrient loss in conventional methods. FYM should not be opened to the sun and should be protected from rain, if possible. Occasional stirring or agitation can provide good aeration during decomposition.

Farm Yard Manure Application

Water Management

Bmps on Water Management are Developed to improve the water use Efficiency in irrigated Cotton.

The water related BMPs include adoption of water conservation techniques like drip irrigation and innovative techniques like paired row planting of cotton (60 cm apart as against 90-120 cm apart) coupled with 120 cm inter spaces between 2 paired rows. Other methods include skip row irrigation and furrow application technique by farmers who otherwise adopt flood irrigation irrespective of row length.

Critical Stages of Moisture Requirement for Cotton

There are many stages during crop cultivation, when moisture is important for its growth. Moisture is critical at the entry level of each stage. If the plant undergoes water stress during these stages, yields are significantly reduced. Following are some points that the extension worker should keep in mind with respect to irrigation:

- If water is available for one irrigation, it should be provided at the flowering stage.
- If water is available for two irrigations, it should be provided at the flowering and boll formation stages.
- If water is available for three irrigations, it should be provided at the seedling, flowering and boll formation stages.
- If water is not limited, irrigation should be provided every 15 days and water stagnation should be avoided.

Water Management Practices in Cotton Cultivation

- Straight Furrow: This is a surface irrigation method provided in between two furrows. Furrows are developed between cotton crop rows during planting and interculture. Water is applied by running small streams in furrows between crop rows. Water infiltrates into the soil and spreads laterally to irrigate the area between the furrows.

- Alternate Irrigation: Under this method, the area in between the rows is irrigated using a combination of flood and furrow irrigation. In case of cotton, irrigation can be provided to alternate rows, by skipping a row to save water usage. This has the potential to save around 35% of irrigation water in comparison to the conventional furrow method in sandy loam soil.

- Paired Row Technique: Paired row technique is an extension of alternate irrigation. It is a method in which additional crop rows are added on both sides of the furrow, besides increasing the ridge spacing, thereby allowing one furrow to irrigate two paired rows. The paired row system of planting cotton crop can save around 30 % of irrigation water with no reduction in the yield as compared to the conventional furrow system.

- Drip Irrigation Method: Application of drip or micro irrigation results in saving water and enhances the water use efficiency of cotton to a great extent. Regulated and slow application of irrigation water through emitters/orifices enables the water to reach the root zone of plants at frequent intervals. Drip irrigation has been found to enhance the yield of cotton by 27%.

Drip Irrigation

Pest and Disease Management

- Ploughing practices: Boll worms during pupal stage hide in debris, stubbles and upper layers of the soil. Summer deep ploughing practices are adopted to expose them to sun light. It helps to expose the pupae of bollworms, which are exposed to sunlight and predation by birds.

- Enrichment of compost to fight soil borne plant diseases: When compost is enriched by mixing and growing natural antagonistic fungi (Triochoderma) and applied as a top dressing at the base of the plants, the practice helps in effective deployment of biocontrol fungus to control wilts and other soil borne diseases, apart from promoting plant growth.

- Trap, border and inter crops: Trap and border crops like marigold, castor, maize, sorghum and pigeon pea are grown in cotton fields to serve as trap crops for insects (mostly bollworms) and to increase the in-field and border diversity to serve as barrier or guard crop for reducing pest dispersal and in some cases to serve as food resources for enhanced multiplication of beneficial insects. Green gram, as an intercrop, not only brings in additional income, but also helps in in-situ conservation of soil moisture and promotes the development of natural enemies of pests thereby reducing the pest incidence on cotton.

Intercropping

Pest	Recommended trap crop (100 plants/acre)
Bollworm, Helicoverpa armigera	Marigold, sunflower, sorghum, pigeonpea
Bollworm & tobacco caterpillar, Spodoptera litura	Castor
Spotted bollworm & Pink bollworm	Okra

Helicoverpa infestation in cotton leaves

Marigold on field bunds can attract Helicoverpa bollworm and castor can attract tobacco caterpillar.

Pest monitoring at weekly intervals can help the farmers identify eggs layed by these insects that are collected and destroyed. Tobacco caterpillar is generally found in groups during the first 5 days after hatching, under infested leaf surfaces, which then show up as papery-white patches due to larval scrapping of leaves and these can be spotted from a distance, manually picked and destroyed. Marigold, castor, bhindi, sunflower should be sown as trap crops and monitored for pest incidence.

- Pheromone traps: Female insects release a kind of scent to attract male insects for mating. Male insects get attracted to this smell and wander to the source of the smell. This chemical attraction is species specific. Synthetic insect specific pheromones are available as lures for deployment in suitable dispensers housed in traps designed for maximum catching efficiency of attracted adult males (moths). The whole apparatus is deployed in the field tied to a wooden stake one foot above the crop canopy. The pheromone trap serves the purpose of tracking the onset of pest activity in the field. Trapped insects during the night are counted the next day and removed after enumeration. Based on the intensity of pest infestation, control measures are adopted. Separate pheromone traps are available for Helicoverpa, tobacco caterpillar and pink bollworm. The farmer, based on his experiences of pest infestation, can opt for specific pheromone traps for estimating the intensity of the pest attack.

- Color traps: Some sucking insect pests get attracted to specific colours. Colored boards/sheets/round tins should be arranged one foot above the crop by applying castor oil on each surface and tied to a stake. Yellow colour bait trap works for trapping whiteflies while white colour boards works for jassids. The insects attracted to the colour boards get stuck to the sticky traps and can be checked for the intensity of the infestation in the field. Periodically the surfaces need to be cleaned and castor oil needs to be applied again. In one acre, 2-3 yellow sticky traps could serve the purpose of monitoring.

- Light Traps: Many insects get attracted to light. Moths, butterflies, adult sucking pests get attracted to light or flame. The light trap idea is based on this behavior. An electric bulb of 100-150 watts along with a water trough below that at one feet distance should be arranged near the electricity source. Insects that get attracted to the light operated during night time fall in the water and get killed when few drops of kerosene/soap powder is added to water. Pests are observed flying in the field in the evening hours between 7-9 pm. Switching off the light beyond this time will not affect beneficial insects.

Light Trap

- Bird Perches: In nature, many birds feed on insect larvae. Erecting T-shaped wooden poles in the field (10/acre) can serve as a perch for birds. These bird landings in the field promote natural pest control. Results are also observed by pouring water in broken mud pots. Some farmers spread rice mixed with turmeric in the field to attract birds. Shrikes, mynas and baya weaver birds land on bird perches. Ideally, four to five bird perches should be arranged one ft above the crop.

Botanical extractions like neem seed kernel extract, vitex decoction, tobacco decoction and neem oil are used for pest control. Through weekly monitoring, the diversity and intensity of the insects can be understood. At times, when there are no beneficial insects in the field and when the conditions are not favourable for them, the pest intensity will be higher.

Neem seed oil extraction

Under favourable conditions sucking pests spread across the entire field in a short time. When the farmer feels the pest situation is out of control, he should prepare the extracts from the available trees in the village.

When adult pests are attracted to light or trapped in pheromone traps, neem sprays and concoctions prepared from five leaves should be sprayed to avoid egg laying by bollworms. These sprays are intended to bring back the natural balance of pests and beneficial insects, but not as an alternative to chemical sprays. Based on the diversity and intensity of pests, extracts should be sprayed to bring back the natural balance. Bio pesticides are also recommended for controlling bollworms.

Harvest and Storage

Better Management Practices also emphasize safe and Clean Picking and storage practices to promote Clean Cotton for Better Prices and Markets.

Picking of Cotton

The time of picking is an important aspect for maintaining seed quality. Picking should only commence when the cotton is fully mature, i.e., when the bolls begin to open. Several pickings may be necessary since bolls ripen every two to three months. Early cotton picking gives slightly better seeds for germination. In addition, the planting seed is best gathered at the peak of the harvest. The cotton picked from late-formed bolls (last pickings) should not be kept for seeds.

Precautions to be taken in picking:

1. Picking should be started when bolls are fully mature.
2. Picking should not be done while the bolls are wet from dew or rain.
3. Bolls spoiled by rains, or damaged by insects, or otherwise damaged, should be picked separately and discarded. Such bolls can be picked 10 to 15 days ahead of the first picking. The damaged bolls should not be picked during normal pickings for seed purposes.
4. Cotton should be clean, with a minimum amount of material such as leaves and plant parts. This reduces the possibility of mechanical damage from excessive machining of the fibre during ginning.
5. Moist cotton should not be picked or stored in any way. At a moisture content of twelve per cent, or more, heating may occur and damage the seed and fibre. Further, damp cotton requires more processing in the gin.
6. The picked cotton, when it is completely dry, should be stored in a dry place and should be covered if not ginned immediately.

References

- ProGuiFlax: nda.agric.za, Retrieved 13 March, 2019
- ProdGuideSunflower: nda.agric.za, Retrieved 3 July, 2019
- How-is-rice-grown, rice-as-a-crop: ricepedia.org, Retrieved 25 June, 2019
- Pearl Millet: daff.gov.za, Retrieved 15 April, 2019
- Cultural Management Practices: icrisat.org, Retrieved 23 January, 2019
- Mustard: purdue.edu, Retrieved 3 May, 2019
- Better-management-practices-for-cotton-cultivation: wwfindia.org, Retrieved 13 February, 2019

Chapter 6

Understanding Farming Systems

A community of separate farm systems which have broadly similar enterprise patterns, household livelihoods and resource bases is known as a farming system. There are various areas of study under farming systems such as plantation agriculture, crop rotation, dry farming and ley farming. The following chapter elucidates these varied processes and mechanisms associated with this area of study.

A farming system is defined as a population of individual farm systems that have broadly similar resource bases, enterprise patterns, household livelihoods and constraints, and for which similar development strategies and interventions would be appropriate. Depending on the scale of the analysis, a farming system can encompass a few dozen or many millions of households.

The classification of the farming systems of developing regions has been based on the following criteria:

- Available natural resource base, including water, land, grazing areas and forest; climate, of which altitude is one important determinant; landscape, including slope; farm size, tenure and organization.
- Dominant pattern of farm activities and household livelihoods, including field crops, livestock, trees, aquaculture, hunting and gathering, processing and off-farm activities; and taking into account the main technologies used, which determine the intensity of production and integration of crops, livestock and other activities.

Advantages of Integrated Farming System

- Higher food production to equate the demand of the exploding population of our nation.
- Increased farm income through proper residue recycling and allied components.
- Sustainable soil fertility and productivity through organic waste recycling.
- Integration of allied activities will result in the availability of nutritious food enriched with protein, carbohydrate, fat, minerals and vitamins.

- Integrated farming will help in environmental protection through effective recycling of waste from animal activities like piggery, poultry and pigeon rearing.
- Reduced production cost of components through input recycling from the by-products of allied enterprises.
- Regular stable income through the products like egg, milk, mushroom, vegetables, honey and silkworm cocoons from the linked activities in integrated farming.
- Inclusion of biogas & agro forestry in integrated farming system will solve the prognosticated energy crisis.
- Cultivation of fodder crops as intercropping and as border cropping will result in the availability of adequate nutritious fodder for animal components like milch cow, goat / sheep, pig and rabbit.
- Firewood and construction wood requirements could be met from the agroforestry system without affecting the natural forest.
- Avoidance of soil loss through erosion by agro-forestry and proper cultivation of each part of land by integrated farming.
- Generation of regular employment for the farm family members of small and marginal farmers.

Components of Integrated Farming System

- Crops, livestock, birds and trees are the major components of any IFS.
- Crop may have subsystem like monocrop, mixed/intercrop, multi-tier crops of cereals, legumes (pulses), oilseeds, forage etc.
- Livestock components may be milch cow, goat, sheep, poultry, bees.
- Tree components may include timer, fuel, fodder and fruit trees.

Factors to be considered:

The following factors have to be considered while selecting IFS in rainfed areas.

Soil types, rainfall and its distribution and length of growing season are the major factors that decide the selection of suitable annual crops, trees and livestock components. The needs and resource base of the farmers also decides the selection of IFS components in any farm.

Suitable Grain Crops

According to soil type we can select suitable crops.

Black Soil

Cereals	Maize
Millets	Sorghum, bajra
Pulses	Greengram, blackgram, redgram, chickpea, soybean, horse gram
Oilseeds	Sunflower, safflower
FIbre	Cotton
Other crops	Coriander, chillies

Red Soil

Millets	Sorghum
Minor Millets	ragi, tenai, samai, pani varagu, varagu
Pulses	Lab- lab, greengram, red gram, soybean, horse gram, cowpea
Oilseeds	Groundnut, castor, sesame

Suitable Forage Crops

Black Soils

Fodder sorghum, fodder bajra, fodder cowpea, desmanthus, Rhodes grass, Mayil kondai pul, Elusine sp., Thomson grass.

Red Soils

Fodder cholam, fodder bajra, Neelakolukattai (Blue Buffel Grass), fodder ragi, Sanku pushpam (Conch flower creeper), fodder cowpea, Muyal Masal (Stylo), siratro, marvel grasses, spear grass, vettiver.

Suitable Tree Species

Tamarind, Simarouba,Vagai (Ladies tongue), Arappu, Kodai vel, A.tortilis, Maan Kathu vel, A.mellifera, Neem, Hardwickia binata, Ber, Indian Gooseberry, Casuarina, Silk cottonetc. are suitable for red gravelly/sandy red loam soils.

Karu vel, A.tortilis, A.albida, Neem, Vagai, Holoptelia integrifolia, Manja neythi, Hibiscus tilifolia, Gmelina arborea, Casuarina, Subabuland Adina cordifolia are suitable for black soils.

Suitable Livestock and Birds

- Goat, sheep, white cattle, black cattle, pigeon, rabbit, quail and poultry.
- Agronomic approaches for increasing overall productivity and sustainability of IFS.
- The various agronomic approaches for increasing the overall productivity and sustainability of IFS:
    - Adoption of improved cropping system according to the rainfall and soil moisture availability.
    - Selection of suitable grain crop species, tree species that supply pods/leaves for a longer period or throughout the year.
    - The surplus fodder leaves, crop residues etc. during the rainy season should be preserved as silage/hay for lean season (summer).

Integrated Farming System with Goats and Sheep

Production of agricultural crops, vary in response to changes of the seasons. In the recent period stable income of agricultural crops has become unstable. Redressing these by integrating crops with agro-based industries like livestock farming is essential. In today's environment, increasing demand for meat in rural and urban areas can help small and part-time farmers earn good profits by trading with the goat and sheep farming. Because of less investment and maintenance costs for shed, it can be done in coordination with agriculture crops.

Sheep farming in Integrated Farming suitable for wetland: In one acre of land goats can grow up to 30 to 35 numbers. The forage crops of cumbu Napier CO_4 (40 cents), Hedge lucerne - (30 cents), fodder sorghum CO.F.S. 29 - (30 cents) these crops can ensure supply of fodder throughout the year. In this CO_4 grass is high forage yielding one. It had high tillering capacity, lean stems with more leaves, easy palatability and not having spines attracts sheep. Further, it has easy digestibility. The intercrop CO_4 variety are fed into small pieces will increase the body growth rate. Buffel grass varieties viz., blue, white and black can withstand the drought and weed infestation. This kind of grass was suitable for pastures. These grasses cultivated with legume fodder such as Stylo in 3:1 proportion in dryland pastures would increase the productivity of goats. Produced forages are cut in to small pieces and fed to the sheeps. Generally, sheep produce three lambs in two years. This will give more profitable than other livestock. Moreover, green fodder given to goats about 2 to 3 kg per day would be sufficient.

- Goat rearing in dryland IFS: In integrated farming, suitable for dryland of 1 hectare land with crop cultivation and goat rearing (20 females: 1 male) by doing this, three times production, net profit and increasing employment chances.

From 20 sheeps we can get 45 lambs in a year. Moreover, from sheep manure we get 200 kg N, 106 kg P and 91 kg K. it also gives 40 to 50 thousand rupees additional income. In our country, sheeps are largely dependent on Grazing land because of this sheep's productivity is low. Solving this problem tree kind fodder leaves, agri related products as daily feed will increase the productivity. Thus groundnut leaves, red gram bran, black gram bran and such as the feeding of the wood leaves like Agathi, Barnyard, neem, Tamarind, Supapul, Desmanthus, Portia, surrogates such as tree leaves Acacia, Kutaivelan, Velikattan and Raintree pods will give required nutrient-rich forage to the sheep. It removes fodder demand, increase meat production leads to getting additional income. During rainy season, sheeps get enough amount of green grass through the pasture. So allow 6 to 8 hours daily for grazing is enough for sheep. During summer (March to August) green grass not available. Therefore, in the Cause of mixed ration (Concentrate) giving to sheep's is necessary. In this concentrated fodder include cereals, cakes, rice or wheat bran, mineral Mixture and normal saline. Usually for sheep's, groundnut cake, sesame or mixed soya cake mixed with mixture fodder may fed to the sheep. During non-availability of mixture fodder, red gram bran, black gram bran, bengal gram brans to be given as fodder.

Agro forestry and goat farming: Practicing agriculture with livestock farming is more profitable than doing agriculture alone. In addition to goat rearing in agriculture, growers growing goats 20 to 30, per year will give minimum 40 to 50 thousand rupees as additional profit. Agroforestry and goat rearing doing together may give many benefits to the farmers.

- Fallow land and dry land may be exclusively used for this project.
- In this method, the amount of water needed is very low.
- In this method, from the goat dung and leaves heap decomposition used as fertilizer for agriculture, the opportunity to increase soil fertility.
- Fifth year onwards, get more profit from well grown tree.
- Above all, farmer will get the employment opportunity throughout the year.

Agroforestry and Kennel Method of Goat Rearing

- New forage variety hedge lucerne in 25 cents and kollukattai grass in 25 cents should be cultivated.
- For an acre based on water facility available at least 25 cents fodder maize 29, 5 cents of land used to build fences and set up kennel method shed in the center.
- In the remaining land, Agathi, Soundal, Gliricidia trees could be planted in the border.
- Three to four months prior to the purchase of goat starts cultivating forage crops and kept ready.
- Goat rearing in kennel method.

Construction of Shed

- 10 to 15 square feet of space is needed for one goat.
- Put shed with coconut palm leaf or tiles set.
- Aluminum plates on one side of the barn to feed and keep.
- Water supply with automatic tool.
- Kennel should be set at a height of about 3 feet above ground level.
- Mixed forages 50g for younger ones, 100 g for grown and 200g for pregnant sheep per day. For goats, requires 1-3 litres of water per day. It also fed with 1 to 2 kg of forage sorghum and 250 to 300 grams of tree leaves per day.
- Integrated sheep rearing: Sheep farming in dry lands under integrated farming system is profitable business. In these lands crop cultivation with fodder trees are fed to the sheep. It has rich in proteins and minerals. These include Agathi, neemSupapaul, Acacia sp, kalyanamurungai, Gliricidia and Raintree are important one. Tree leaves are rich source of nutrients than other fodder. Instead of giving tree fodder alone it can be mixed with cereal or legume fodder will reduce the feed cost of fodder sorghum and increase the productivity of sheep. It's not just giving maravakait tivanankalait separately mixed with cereal grains or forages pulvakait, sheep concentrate mixture of corn, cut through the green, can increase productivity. Sheep farming income depends on the ability of production of lamb. To make the sheep farming profitable one breeding management practices to be followed in right time. Goats are usually produces their offspring's around the year and also get into lactation 60 to 90 days after give berth.

Goats become enter into estrus period during March to April give birth to lamb in July to August. From the beginning of June, due to start of south-west monsoon both sheep and lamb got good fodder. The sheep had good milk production capacity increase the growth of lamb.

During the months of July and August those sheep eve lambs in November and December. In some areas there is severe cold and snowfall caused growth and health of lambs. Similarly, in places that are dependent on rainfall areas fodder shortages are likely to occur. Separated from the mother the offspring not provided sufficient amount fodder will cause growth of the younger ones. The increase in mortality of the offspring. Rainfed crops harvested in the months of December and January may be used as feed for mother to increase the milk yielding capacity. if they do get to keep the mother sheep byproducts of agricultural products can be corrected by maintaining milk production.

Methods to Increase Fodder Production

- Grow high yielding varieties and technologies.
- Area for forage production is very low. So we can increase the size of the area planted forages as possible.
- Legume fodder may intercropped with other crops to increase the production.
- Social forestry and Agroforestry system forage crops can be grown. For example, between the trees supapaul kolukattai grass and stylo can be grown in 3: 1 proportion.
- Fallow land, lake, forage or fodder trees can be grown in places like roadside land.
- Agricultural lands, fruit tree mango, sapota, guava, lemon, coconut and tamarind intercrop forages grown between the trees.
- Grow good grass pastures in poor condition land Marvel grass, kolukattai grass and legume fodder like Stylo and ciratro can be sown.

Disadvantages of Integrated Farming System

- When planning to receive double benefits, one should also be prepared to bear double responsibility as all the different spheres the farmer chooses to combine will have their own specific features.
- It often happens that fish farming, cattle rearing and plants farming contradict each other. For example, plants should not be treated with substances that can harm animals.
- It is necessary to thoroughly understand what you are going into. For example,

if you plan to improve your yield with the help of manure, it is important to know how much is needed to avoid overdoing it, which will bring more harm than good. To prevent pollution, you need to be good at management and care.

- Scientists argue about the benefits of integrated farming. The process must be well thought through, down to the smallest details, so as to not contradict the norms of public health. Some experts say that combining birds, pigs and fish can cause diseases, such as influenza. It is known that human and avian influenza can be mixed in the body of a pig and new deadly viruses can arise in the process of mutation. There is no exact confirmation that this would happen, however, it is better to be cautious. As a security measure, farmers must abstain from combining pigs with birds.

Plantation Agriculture

Plantation agriculture is a form of commercial farming where crops are grown for profit. Large land areas are needed for this type of agriculture. Countries that have plantation agriculture usually experience high annual temperatures and receive high annual rainfall. Plantations are mainly found in countries that have a tropical climate.

The plantation agriculture is practiced mainly in the tropical areas to grow cash crops. It is a specialized commercial cultivation of cash crops on estates or plantations. Some of the important plantation crops are rubber, oil-palm, cotton and copra, beverages like tea, coffee and cocoa, fruits like pineapple, banana, sugarcane, hemp and jute.

The farming in plantation is executed with specialized skill and, wherever, possible with the application of machinery, fertilizers, Weedicides, insecticides and pesticides. It aims at high yields, high quality production and a large output, most of which is exported.

The final products, whether sheet rubber, palm-oil or tea, have to be carefully processed and standardized to meet the world demand and specifications. The competition in such production is very keen. Large estates are; therefore, better able to meet these demands than individual, small holders and farmers.

In the tropics and subtropics where plantation agriculture is practiced, the use of white labour was known impossible during the eighteenth and nineteenth centuries for climatic and health reasons. All parts of the tropical climate are, however, not equally well suited to the development of plantation agriculture. Some areas have very heavy and continuous rainfall, others have very little precipitation.

The excessive amount of water in the soil alters various chemical and biological processes, limiting the amount of oxygen and increasing the formation compounds that are toxic to plant roots. On the other hand, drought and scanty rainfall results into withering of crops if moisture to the crop is not supplied artificially.

Again, temperature conditions are not always adapted to the growing of the desired products nor are they favourable for man. Further, climatic hazards, strong winds, topography, drainage, soil, vegetation conditions often handicap or may even prevent the development and establishment of plantation estates.

Plantation Agriculture: Location and Characteristics

Location

Plantation agriculture is confined within tropical areas, i.e., both sides of the equator. Plantations exist on every continent possessing a tropical climate. The plantation system however is considerably older in tropical America than in Asia and Africa. The tropical areas of Latin America, Asia and Africa are the areas where plantation agriculture has been developed.

Some of the important plantation farming are coffee plantation in Brazil, Paraguay and Bolivia, Tanzania, Kenya; sugarcane plantation in Cuba, Brazil, Peru, Puerto Rico and Philippines; tea plantation in India, Sri Lanka, Indonesia; cocoa farming in West Indies, Ecuador, Brazil, Nigeria, Ghana; rubber plantation in Malaysia, Indonesia, Thailand, Sri Lanka, Cambodia, Myanmar, India. banana plantation in Mexico, Jamaica, Columbia, Brazil, Panama and Costa Rico.

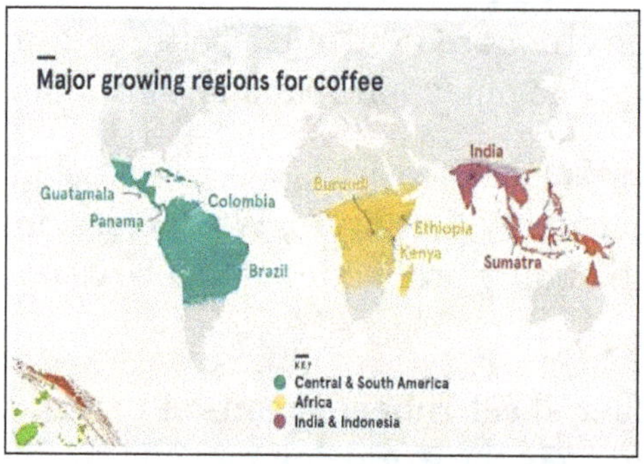

Major plantation areas of the world

Characteristics

Plantation agriculture is different than other forms of agriculture. The characteristic features of plantation agriculture are as follows:

(i) Plantation agriculture is highly sophisticated and scientific methods are used for large-scale production.

(ii) There is specialization of single crop in plantation agriculture, e.g., coffee in Brazil, tea in India, rubber in Malaysia, etc.

(iii) Estate farming: Plantation crops are usually raised on large estates, of more than 40 hectares (100 acres) each, though the success of such crops has often encouraged other farmers to grow them so that small holdings exist side by side with the large estates.

- In some parts of West Africa small holdings are, however, more important than estates and small holding rubber production in Malaysia now exceeds that from estates. Seedlings are first grown in nursery seed-beds and then transplanted in neat rows, well-spaced and regularly weeded, on the estates.

(iv) Foreign ownership and local labour:

- Most of the largest estates are owned by Europeans. For example, most Malaysian rubber estates were originally in the hands of British companies with their head offices in London, and were managed and supervised by Englishmen.
- Since independence, however, there has been an increase in local ownership. The tapping and processing of the rubber is done entirely by local people or by immigrant labourers from southern India.
- The British also established large tea gardens in India and Sri Lanka and banana and sugar plantations in the West Indies. The French have established cocoa and coffee plantations in West Africa, e.g., in Cameroun and Ivory Coast, but Ghana and Nigeria have fewer estates.
- The Dutch once monopolized the sugarcane plantations in Indonesia, especially in Java. Spanish and American capitalists invested heavily in coconut, abaca and sugar plantations in the Philippines.
- Some coffee fazendas in Brazil are still in the hands of the Portuguese, though most of them belong to wealthy Brazilians, and there is also some Spanish, American and Canadian interest in plantation agriculture in tropical America. Sugar plantations in Queensland, Australia, are unusual as they employ white labour.

(v) Farming in estates is scientifically managed:

- Work in estates is executed with specialised skill, and wherever possible with the application of machinery and fertilizers. It aims at high yields, high quality production and a large output, most of which is exported.
- The final products, whether sheet rubber, palm oil or tea, have to be carefully processed and standardised to meet world demand and specification.

(vi) Heavy capital outlay:

- To initiate and to maintain a tropical plantation, large sums of money are required. As many plantations are located in previously undeveloped or sparsely populated areas, far from urban centers, a minimum network of communication by road or rail has to be developed first, either with or without government assistance.

- This is a very expensive undertaking especially under tropical conditions where maintenance costs are also high. Plantations are manned more and more by local staff, however, less administrative expenses are incurred, but the overall production cost is still high, especially where labour is highly unionised as in Indian and Sri Lankan tea estates.

(vii) Plantation agriculture is an export-oriented agriculture:

- Nearly the entire product of plantation agriculture is generally exported to the international market. Therefore, it requires a better communication network, packaging and processing facilities.

(viii) Unlike other forms of agriculture, plantations are well-planned not only in terms of field structure but also having facilities like residence, transport, hospitals, retail markets, etc., for workers and associated people and their families.

(ix) Plantation agriculture often encourages migration from other countries. In colonial period thousands of people migrated from one place to another for work. In this way, cultural exchange occurs.

(x) Most of the crops grown in plantation agriculture have a life cycle of more than two years. Natural rubber, coconuts, oil palm, tea, cocoa, and coffee are all tree crops and take years to mature, but afterwards they are productive for long periods.

Crop Rotation

Crop rotation refers to the cultivation of different crops on a particular piece of land over time. The succession of crops to be grown is carefully designed to ensure soil nutrients are sustained, pest populations are controlled, weeds are suppressed and soil health is built.

A crop rotation will cycle through cash crops (such as vegetables), cover crops (grasses and cereals) and green manures (often legumes). The exact sequence of crops will vary depending on local circumstances, with the critical design element being an understanding what each crop contributes and takes from the soil. For instance, nitrogen depleting crop should be preceded by a nitrogen fixing crop.

The central idea is to have the crops themselves sustain soil health, rather than planting the same crop year in, year out, and then repairing soil health through fertilisers, pesticides and herbicides.

Advantages of Crop Rotation

A well designed crop rotation makes land both more productive and more environmentally sustainable. It improves the financial viability of a farm by increasing productivity whilst reducing chemical input costs. Key advantages of crop rotation are:

- Improved soil fertility and structure.
- Disease control.
- Pest control.
- Weed control.
- Increased Soil Organic Matter.
- Erosion control.
- Improved biodiversity.
- Increased yield.
- Reduced commercial risk.

PRI Zaytuna Farm

Improved Soil Fertility and Structure

Crop rotation improves the physical and chemical conditions of soil and thus improves the overall fertility. Nitrogen-fixing legumes such as soybeans and alfalfa in crop rotations fix atmospheric nitrogen into the soil through root nodules. This nitrogen is then available for subsequent crops.

Deep rooted cover crops can draw up nutrients such as potassium and phosphorus from deep in the soil profile, making these nutrients available for subsequent shallow rooted cash crops.

Growing a hay crop in a rotation can result in improved tilth and bulk density. When a hay crop is ploughed in, the soil will be loose and have a good granular structure and tilth. These improved properties result from the soil being protected from raindrops, the network of fine roots in the soil, and the formation of humus from decomposing plant roots."

Disease Control

Crop rotation helps to control common root and stem diseases that affect row crops. Crop rotation is highly effective against diseases whose pathogens have a small host range and require soil or crop residue to overwinter. For such diseases, rotating a non-host crop immediately after a host crop prevents the pathogen from reproducing. The pathogen inoculum, ordinarily preserved in crop debris, does not have the necessary conditions for its survival and the disease spread is controlled. For example, soybean cyst nematode populations can be cut in half by rotating soybean with wheat and corn.

In the absence of crop rotation, growing the same crop on a particular parcel of land year after year gives pathogens continued optimal conditions, and their population will increase rapidly.

Pest Control

Crop rotation can be used as a tool to manage those insects which are non-mobile, whose larvae or eggs overwinter in soil and which have a narrow range of crops to feed on.

For instance, corn rootworms can be managed effectively with crop rotation. These insects lay eggs in the corn fields they live on, and emerge to damage subsequent crops. Rotating a non-host crop immediately after a corn crop means that emerging larvae starve due to scarcity of food.

Weed Control

PRI Zaytuna Farm

Including cover crops into crop rotation systems provides greater competitions to the weeds for their basic needs such as nutrients, space and light. Cover crops ultimately crowd out the weeds, slowing down weed growth and proliferation for a reduced weed population in subsequent crops.

Increased Soil Organic Matter

Crop rotation will add more crop residues, green manures and other plant debris to the soil. Crop rotation also requires less intensive tillage, which means that soil organic matter does not degrade as quickly.

Increased soil organic matter improves soil infiltration and water holding capacity, which enables water to be absorbed into the soil. Furthermore, increased of soil organic matter improves overall soil structure and the chemical and biological properties of the soil.

Erosion Control

Crop rotation helps control the erosion of soil from water and wind by improving the soil structure and reducing the amount of soil that is exposed to water and wind. Crop rotation also supports reduced or no-till farming, which ensures even better protection against erosion.

PRI Zaytuna Farm main crop

Cover crops are effective in reducing raindrop impact, reducing sediment detachment and transport, slowing surface runoff, and so reducing soil loss.

To maximise the results of erosion control efforts, the crop rotation should be designed to reflect climatic conditions. For example, rigid crop rotations give good crop growth along with effective soil cover under consistent climatic conditions of fairly predictable annual rainfall and temperature. More flexible rotations should be used in regions that are susceptible to unseasonal rains or drought.

Improved Biodiversity

Crop rotation helps improve soil biodiversity by changing crop residue and rooting pattens. Different crops benefit different species, and so a range of crops will lead to a more diverse and healthy soil microbial community. Similarly, the microbial community is supported by rotating crops with a high carbon to nitrogen ratio (such as corn) with low carbon to nitrogen ratio crops (such as soybeans).

Increased Yield

Crop rotation can help increase yield. Corn and soybean that is rotated with another crop yields 10% more than when the same crop is grown continuously". The increased yield is the result of all of the individual soil and plant health benefits from crop rotation.

Reduced Commercial Risk

Different crops have resistance capacities against different adverse climatic conditions. For instance, some crops have good tolerance against flooding conditions while some others have improved drought resistance. Growing different crops in rotation minimizes the impact of crop failure due to adverse weather.

Crop rotation also requires growing and harvesting crops at different times, helping farmers to spread their workload evenly and allow them to cultivate more land with same amount of equipment and labour.

Dry Farming

Dry Farming, also called Dryland Farming, is the cultivation of crops without irrigation in regions of limited moisture, typically less than 20 inches (50 centimetres) of precipitation annually. Dry farming depends upon efficient storage of the limited moisture in the soil and the selection of crops and growing methods that make the best use of this moisture. Tilling the land shortly after harvest and keeping it free from weeds are typical methods, but in certain latitudes stubble is left in the fields after harvest to trap snow. Moisture control during crop growing consists largely of destruction of weeds and prevention of runoff. The ideal soil surface is free of weeds but has enough clods or dead vegetable matter to hinder runoff and prevent erosion.

Crops adapted to dry farming may be either drought resistant or drought evasive. Drought-resistant crops, such as sorghum, are able to reduce transpiration (emission of moisture) and may nearly cease growing during periods of moisture shortage, resuming growth when conditions again become favourable. Drought-evasive crops achieve their main growth during times of year when heat and drought conditions are not severe. Crops adapted to dry farming are usually smaller and quicker to mature than those grown under more humid conditions and are usually allotted more space.

Features

Locations

Fields in the Palouse

Dryland farming is used in the Great Plains, the Palouse plateau of Eastern Washington, and other arid regions of North America such as in the Southwestern United States and Mexico the Middle East and in other grain growing regions such as the steppes of Eurasia and Argentina. Dryland farming was introduced to southern Russia and Ukraine by Ukrainian Mennonites under the influence of Johann Cornies, making the region the breadbasket of Europe. In Australia, it is widely practiced in all states but the Northern Territory.

Crops

Dryland farmed crops may include winter wheat, corn, beans, sunflowers or even watermelon. Successful dryland farming is possible with as little as 230 millimetres (9 in) of precipitation a year; higher rainfall increases the variety of crops. Native American tribes in the arid Southwest survived for thousands of years on dryland farming in areas with less than 250 millimetres (10 in) of rain. The choice of crop is influenced by the timing of the predominant rainfall in relation to the seasons. For example, winter wheat is more suited to regions with higher winter rainfall while areas with summer wet seasons may be more suited to summer growing crops such as sorghum, sunflowers or cotton.

Process

Dryland farming has evolved as a set of techniques and management practices used by farmers to continually adapt to the presence or lack of moisture in a given crop cycle. In marginal regions, a farmer should be financially able to survive occasional crop failures, perhaps for several years in succession. Survival as a dryland farmer requires careful husbandry of the moisture available for the crop and aggressive management of expenses to minimize losses in poor years. Dryland farming involves the constant assessing of the amount of moisture present or lacking for any given crop cycle and planning accordingly. Dryland farmers know that to be financially successful they have to be aggressive during the good years in order to offset the dry years.

Dryland farming caused a large dust storm in parts of Eastern Washington

Dryland farming is dependent on natural rainfall, which can leave the ground vulnerable to dust storms, particularly if poor farming techniques are used or if the storms strike at a particularly vulnerable time. The fact that a fallow period must be included in the crop rotation means that fields cannot always be protected by a cover crop, which might otherwise offer protection against erosion.

Some of the theories of dryland farming developed in the late 19th and early 20th centuries claimed to be scientific but were in reality pseudoscientific and did not stand up to empirical testing. For example, it was alleged that tillage would seal in moisture, but such "dust mulching" ideas are based on what people imagine *should* happen, or have been told, rather than what testing actually confirms.

Capturing and conservation of moisture – In regions such as Eastern Washington, the average annual precipitation available to a dryland farm may be as little as 220 millimetres (8.5 in). Consequently, moisture must be captured until the crop can utilize it. Techniques include summer fallow rotation (in which one crop is grown on two seasons' precipitation, leaving standing stubble and crop residue to trap snow), and preventing runoff by terracing fields.

"Terracing" is also practiced by farmers on a smaller scale by laying out the direction of furrows to slow water runoff downhill, usually by plowing along either contours or keylines. Moisture can be conserved by eliminating weeds and leaving crop residue to shade the soil.

Effective use of available moisture – Once moisture is available for the crop to use, it must be used as effectively as possible. Seed planting depth and timing are carefully considered to place the seed at a depth at which sufficient moisture exists, or where it will exist when seasonal precipitation falls. Farmers tend to use crop varieties which are drought and heat-stress tolerant (even lower-yielding varieties). Thus the likelihood of a successful crop is hedged if seasonal precipitation fails.

Soil conservation – The nature of dryland farming makes it particularly susceptible to erosion, especially wind erosion. Some techniques for conserving soil moisture (such as frequent tillage to kill weeds) are at odds with techniques for conserving topsoil. Since healthy topsoil is critical to sustainable dryland agriculture, its preservation is generally considered the most important long-term goal of a dryland farming operation. Erosion control techniques such as windbreaks, reduced tillage or no-till, spreading straw (or other mulch on particularly susceptible ground), and strip farming are used to minimize topsoil loss.

Control of input costs – Dryland farming is practiced in regions inherently marginal for non-irrigated agriculture. Because of this, there is an increased risk of crop failure and poor yields which may occur in a dry year (regardless of money or effort expended). Dryland farmers must evaluate the potential yield of a crop constantly throughout the growing season and be prepared to decrease inputs to the crop such as fertilizer and weed control if it appears that it is likely to have a poor yield due to insufficient moisture. Conversely, in years when moisture is abundant, farmers may increase their input efforts and budget to maximize yields and to offset poor harvests.

Dry Farming

Locations

Dry farming may be practiced in areas that have significant annual rainfall during a wet season, often in the winter. Crops are cultivated during the subsequent dry season, using practices that make use of the stored moisture in the soil. California, Colorado, Oregon and Wyoming, in the United States, are a few states where dry farming is practiced for a variety of crops.

Crops

Dry farmed crops may include grapes, tomatoes, pumpkins, beans, and other summer crops. These crops grow using the winter water stored in the soil, rather than depending on rainfall during the growing season.

Process

Dry farming depends on making the best use of the "bank" of soil moisture that was created by winter rainfall. Some dry farming practices include:

- Wider than normal spacing, to provide a larger bank of moisture for each plant.
- Controlled Traffic.

- Minimal tilling of land.
- Strict weed control, to ensure that weeds do not consume soil moisture needed by the cultivated plants.
- Cultivation of soil to produce a "dust mulch", thought to prevent the loss of water through capillary action. This practice is controversial, and is not universally advocated.
- Selection of crops and cultivars suited for dry farming practices.

Economic Importance

Dryland farming is highly important to ensure the economic stability of a region or nation with arid lands. In the absence of this farming practice, vast tracts of lands in the world would be left barren and unproductive. Even though dryland farming takes a lot of financial investment and hard work to be established, and crop yields are generally comparatively lower, without this form of agriculture the populations residing in the arid areas of the world would have to be completely dependent on external sources of food to meet their dietary needs. This would adversely impact a nation's economy as self-sufficiency, in terms of production of food grains to feed the country's population, would be lost.

Ley Farming

This system involves the use of various annual and perennial crops (grasses, legumes), grown in long-term and short-term, to manage the soil. They alternate and combine these crops according to their compatibility and growth duration to give optimal desired effects.

Advantages

Improvement of Nitrogen Content for Subsequent Plantations

Legumes improve the fertility of the soil, the quality of the grain and increase the volume of crops. Most of the nitrogen in the soil gets used up by grains, so the alteration with legume crops gives an extremely positive effect.

Effective fixation of nitrogen is most often achieved in mixed legume-grass areas. This element is well captured by perennial herbs. Nitrogen is mainly formed under the root zone and then returns to the surface to interact with crops.

Restoration of Soil and Organic Matter Structure

Active farming deprives the soil of many important substances, and continuous cultivation reduces soil structure and biological activities, reducing the rate at which crops grow.

Ley farming system is an effective tool for soil enrichment by organic material. Legume pastures can't produce the desired effect if used separately; in this case, grass can become an excellent helper, since it promotes the production of organic materials.

It has been scientifically proven that meadow grasses increase soil organic matter by about 0.1% per year. Organic substances accumulate rather slowly so long pasture phases may be needed. This process flows faster on the surface layers.

There have been situations when 5-year use of grass led to a 4 times increase in infiltration of the soil; nevertheless, these advantages are quickly lost if the work on enriching the soil is discontinued. Although it is believed that regular alternation according to the rules of the ley farming system can restore and maintain the soil.

Controlling the Growth of Weeds

Another important advantage of ley farming is the management of weeds since cattles that eat unwanted plants can graze on these pastures. This approach is especially effective in combating weeds annually. The situation gets more complicated when it comes to perennial plants.

Modern agricultural management systems allow effective control from mechanical, chemical and economic points of view. Thanks to this, the adaptability of work of many farmers becomes greatly improved.

Preventing Deep Drainage and Erosion

There is a significant improvement in deepening rates obtained from ley farming in comparison with conventional farming systems. It also reduces erosion during cultivation. On perennial pastures, the root system develops quite well, which provides access to water and nutrients at a considerable depth. As a result, the top layer becomes dry, and the loss of water and useful substances during drainage is prevented. Shallow soil is quite unstable as it does not hold water well, so ley pastures are able to become good tools for reducing drainage.

Improving Livestock Efficiency

Ley pastures are also able to contribute to the improvement of animal husbandry, to provide livestock with high-quality fodder and stimulate the rapid growth of animals. Legumes have the required properties to ensure all these advantages.

Scientific research has proven that the quality of milk can be improved by ley farming. Alternating grass with legumes comes with more advantages than the use of regular grass only. During the grazing season, animals receive high nutritional support so farmers can obtain milk of better quality.

Ley farming also gives farmland owners complete independence since they do not have to buy food for their cattles from other people; they get to know for sure that the food for their livestock is of high quality and will give optimal results.

Ley farming is a balanced system that unites the maximum potential of legumes, grass, and animals. This approach works perfectly in the interests of consumers. Milk production and crop growing are inextricably linked, complement and support each other without emptying the land or reducing fertility.

Natural Fertilizers

It is extremely important that farmers do not have to spend money on expensive fertilizers because the favorable state of the soil gets maintained naturally. They can take advantage of the excellent manure substitute, rather than distribute the standard fertilizer over a vast area of the field, saving time and energy. The grass sod acts well with the feces of animals, without leading to the appearance of harmful by-products.

Disadvantages

To achieve successful development of this method, it is necessary to spend considerable time and invest lots of money.

Problems could also arise due to improper preparation of the seeds, their low quality or unfavorable weather conditions. Success is much more likely to be achieved if one follows the instructions strictly and carry out all the steps as precisely as possible.

It is necessary to remove all modern forms of plant cultivation from the soil before sowing new under the ley farming system, however, some pastures can be difficult to clean. If this is not done, complications could arise that will interfere with the proper development of plants.

Some crops extract water from the depth of the soil, so it can be difficult to enrich them with all the necessary substances before sowing. Sometimes, in order to ensure a good future harvest, it is required to observe a certain interval before removing and sowing plants.

Slash and Burn Cultivation

Slash and burn agriculture is the process of cutting down the vegetation in a particular plot of land, setting fire to the remaining foliage, and using the ashes to provide nutrients to the soil for the use of planting food crops.

The cleared area following slash and burn, also known as swidden, is used for a relatively short period of time, and then left alone for a longer period of time so that vegetation can grow again. For this reason, this type of agriculture is also known as shifting cultivation.

Steps of Slash and Burn

Generally, the following steps are taken in slash and burn agriculture:

- Prepare the field by cutting down vegetation; plants that provide food or timber may be left standing.
- The downed vegetation is allowed to dry until just before the rainiest part of the year to ensure an effective burn.
- The plot of land is burned to remove vegetation, drive away pests, and provide a burst of nutrients for planting.
- Planting is done directly in the ashes left after the burn.

Cultivation (the preparation of land for planting crops) on the plot is done for a few years until the fertility of the formerly burned land is reduced. The plot is left alone for longer than it was cultivated, sometimes up to 10 or more years, to allow wild vegetation to grow on the plot of land. When vegetation has grown again, the slash and burn process may be repeated.

Geography of Slash and Burn Agriculture

Slash and burn agriculture is most often practiced in places where open land for farming is not readily available because of dense vegetation.

Slash and burn is a method of agriculture primarily used by tribal communities for subsistence farming (farming to survive). Humans have practiced this method for about 12,000 years, ever since the transition known as the Neolithic Revolution, the time when humans stopped hunting and gathering and started to stay put and grow crops. Today, between 200 and 500 million people, or up to 7% of the world's population, uses slash and burn agriculture.

When used properly, slash and burn agriculture provides communities with a source of food and income. Slash and burn allow people to farm in places where it usually is not possible because of dense vegetation, soil infertility, low soil nutrient content, uncontrollable pests, or other reasons.

Negative Aspects of Slash and Burn

Many critics claim that slash and burn agriculture contributes to a number of reoccurring problems specific to the environment. They include:

- Deforestation: When practiced by large populations, or when fields are not given sufficient time for vegetation to grow back, there is a temporary or permanent loss of forest cover.

- Erosion: When fields are slashed, burned, and cultivated next to each other in rapid succession, roots and temporary water storages are lost and unable to prevent nutrients from leaving the area permanently.
- Nutrient Loss: For the same reasons, fields may gradually lose the fertility they once had. The result may be desertification, a situation in which land is infertile and unable to support the growth of any kind.
- Biodiversity Loss: When plots of land area are cleared, the various plants and animals that lived there are swept away. If a particular area is the only one that holds a particular species, slashing and burning could result in extinction for that species. Because slash and burn agriculture is often practiced in tropical regions where biodiversity is extremely high, endangerment and extinction may be magnified.

The negative aspects above are interconnected, and when one happens, typically another happens also. These issues may come about because of irresponsible practices of slash and burn agriculture by a large number of people. Knowledge of the ecosystem of the area and agricultural skills could prove very helpful in the safe, sustainable use of slash and burn agriculture.

References

- Agri-majorareas-ifs, agriculture: tnau.ac.in, Retrieved 7 March, 2019
- Advantages-disadvantages-integrated-farming-system: legit.ng, Retrieved 22 May, 2019
- Plantation-agriculture: yourarticlelibrary.com, Retrieved 25 April, 2019
- Plantation-agriculture-location-and-characteristics-with-area-maps: yourarticlelibrary.com, Retrieved 15 February, 2019
- Crop-rotation-a-vital-component-of-organic-farming: permaculturenews.org, Retrieved 9 July, 2019
- Dry-farming: britannica.com, Retrieved 12 January, 2019
- What-is-dryland-agriculture: worldatlas.com, Retrieved 21 May, 2019
- Malcolm, Bill; Sale, Peter"; Egan, Adrian (1996). Agriculture in Australia – An Introduction. Australia: Oxford University Press. ISBN 0-19-553695-9
- Ley-farming-advantages-disadvantages: legit.ng, Retrieved 6 February, 2019
- Slash-and-burn-agriculture: thoughtco.com, Retrieved 19 June, 2019

Permissions

All chapters in this book are published with permission under the Creative Commons Attribution Share Alike License or equivalent. Every chapter published in this book has been scrutinized by our experts. Their significance has been extensively debated. The topics covered herein carry significant information for a comprehensive understanding. They may even be implemented as practical applications or may be referred to as a beginning point for further studies.

We would like to thank the editorial team for lending their expertise to make the book truly unique. They have played a crucial role in the development of this book. Without their invaluable contributions this book wouldn't have been possible. They have made vital efforts to compile up to date information on the varied aspects of this subject to make this book a valuable addition to the collection of many professionals and students.

This book was conceptualized with the vision of imparting up-to-date and integrated information in this field. To ensure the same, a matchless editorial board was set up. Every individual on the board went through rigorous rounds of assessment to prove their worth. After which they invested a large part of their time researching and compiling the most relevant data for our readers.

The editorial board has been involved in producing this book since its inception. They have spent rigorous hours researching and exploring the diverse topics which have resulted in the successful publishing of this book. They have passed on their knowledge of decades through this book. To expedite this challenging task, the publisher supported the team at every step. A small team of assistant editors was also appointed to further simplify the editing procedure and attain best results for the readers.

Apart from the editorial board, the designing team has also invested a significant amount of their time in understanding the subject and creating the most relevant covers. They scrutinized every image to scout for the most suitable representation of the subject and create an appropriate cover for the book.

The publishing team has been an ardent support to the editorial, designing and production team. Their endless efforts to recruit the best for this project, has resulted in the accomplishment of this book. They are a veteran in the field of academics and their pool of knowledge is as vast as their experience in printing. Their expertise and guidance has proved useful at every step. Their uncompromising quality standards have made this book an exceptional effort. Their encouragement from time to time has been an inspiration for everyone.

The publisher and the editorial board hope that this book will prove to be a valuable piece of knowledge for students, practitioners and scholars across the globe.

Index

www.ingramcontent.com/pod-product-compliance
Lightning Source LLC
LaVergne TN
LVHW080633210926
848485LV00003B/257